India and the IT Revolution

India and the IT Revolution

Networks of Global Culture

Anna Greenspan

palgrave
macmillan

First published 2004 by
PALGRAVE MACMILLAN
Houndmills, Basingstoke, Hampshire RG21 6XS and
175 Fifth Avenue, New York, N.Y. 10010
Companies and representatives throughout the world

PALGRAVE MACMILLAN is the global academic imprint of the Palgrave
Macmillan division of St. Martin's Press, LLC and of Palgrave Macmillan Ltd.
Macmillan® is a registered trademark in the United States, United Kingdom
and other countries. Palgrave is a registered trademark in the European
Union and other countries.

ISBN 1–4039–3943–8 hardback

This book is printed on paper suitable for recycling and made from fully
managed and sustained forest sources.

A catalogue record for this book is available from the British Library.

Library of Congress Cataloging-in-Publication Data
Greenspan, Anna.
   India and the IT revolution : networks of global culture / Anna Greenspan.
      p.   cm.
   Includes bibliographical references and index.
   ISBN 1–4039–3943–8 (cloth)
   1. Information technology—Social aspects—India.   2. High
technology—India.   3. Globalization.   I. Title.

   T58.5.G73 2004
   303.48′33′0954—dc22
                                                              2004052304

10   9   8   7   6   5   4   3   2   1
13 12 11 10 09 08 07 06 05 04

Transferred to digital printing 2005

To the creativity, dynamism and innovation of Indian cyberculture, which inspired these cheers from the sidelines.

It was software in cyberspace. There was no system core.

– *Terminator 3: Rise of the Machines,*

A network swarm is all edges.

– Kevin Kelly, *Out of Control*

Contents

Acknowledgments

This book was made possible by two generous research grants. The first was from the Shastri Indo-Canadian Institute, which enabled me – at the start of my project – to spend close to a year in India, traveling to the main IT centers and to some of the country's 'backwaters' to get a first-hand look at how cyberspace was impacting India, and vice versa. The second was from Canada's Social Science and Humanities Research Council, which gave me a two-year postdoctoral fellowship. This allowed me to do research in Silicon Valley, to return to India to get a sense of how the IT industry was surviving in the post-Y2K environment, and to complete a first draft of the book.

My postdoctoral fellowship research was conducted in conjunction with the multimedia program at McMaster University, Canada. Everyone at the university was helpful and welcoming, but I would like to especially acknowledge Dr Geoffrey Rockwell, the director of the multimedia program, who has been consistently supportive of my work.

Many people in both India and America, were kind enough to enrich this book by sharing their experience and insights. These include, in no particular order: from the 'Software Technology Parks of India' S.N. Zindal (in New Delhi), Col. Vijay Kumar and E. Manoj Kumar (in Hyderabad), Sushil K. Gupta (in Pune), and Manas Patnaik (in Bhubaneshwar); from 'TiE', Vish Mishra, Kanwal Rekhi, Kailash Joshi and Rajiv Mathur; software engineer Shashank Gupta; Madanmohan Rao and Osama Manzar, editors at 'Inomy'; IIT Professors Prajit K. Basu, Rukmin Bhayanair and Anand Patwardhan; Rituraj Nath and Sunil Mehta at Nasscom; Vinay Shenoy at Philips; Sumer Shankardass at WNS; Ravi Sundarum from the Centre for Developing Societies, New Delhi; Ashish Sen and Seema B. Nair of 'Voices'; Animesh Thakur and Arun Kumar of Hero Mindmine; Brian Carvalho at ICICI OneSource; Anurag Behar, Ranjan Acharya and Sandhya Ranjit at Wipro; Ravindra Walters at NeoIT; and S. Hariharan at L&T Infocity Limited.

In addition to these I am also thankful to the numerous people at the Hyderabad IT Forum (Hyderabad, 2003); India Internet World (New Delhi, 1999); TiEcon Silicon Valley (Silicon Valley, 2002); TiEcon New Delhi (New Delhi, 2003); and Nasscom India Leadership Forum (Mumbai, 2003), who responded to my questions with both patience and enthusiasm.

My research trips to India were greatly facilitated and made much more pleasurable by my hosts and friends there. I am especially grateful to Alka Amin and her family in New Delhi for making me feel at home, Robin Das for his intense conversations and his family in Orissa for saving me from the supercyclone, Nibha Joshi for sharing her great wisdom, Ravi Rath for all his help in Bhubaneshwar, and Rex van der Spuy for opening up his home to us in Bangalore.

All these people are connected, in one way or another, to Deepti Gupta, whose friendship is my strongest link to India. Deepti has, among other things, accompanied me to Orissa, found me an office space in New Delhi, and shared her apartment, even her room, for close to two months. Deepti and I have spent many hours sipping chai and discussing the future of India and I look forward to many more.

Laura Turcotte, my sister – by destiny if not by blood – was kind enough to proofread both the book proposal and the entire book. All the commas are hers.

I also wish to thank Julia Teng who hosted me in San Fransisco, Beth Chichakian who frequently rescued me from my computer, Michelle Murphy who has taught me many things, not least of which is how to get funding, Sheila Greenspan who accompanied me on what was without doubt my smoothest trip to India, and Louis Greenspan who I always turn to for advice.

My brother Jeremy Greenspan is a talented musician who is growing increasingly famous, so in my own best interest I should thank him here.

Nick Land is not listed as a co-author but he has functioned as practically that. He helped to formulate the first basic ideas in the sweltering heat of Orissa, accompanied me on interviews in Hyderabad and Bangalore, edited the first draft in Canada, and proofread the final manuscript in Shanghai. For this, and for much else, I am both lucky and deeply grateful.

This book owes most to my two greatest influences – my family and the Cybernetic culture research unit (Ccru).

Notes

The value of the rupee fluctuates. In March 2004, 1 US dollar was worth 45 rupees.

Indians use the terms 'lakh' and 'crore' when counting high numbers. One lakh equals one hundred thousand and one crore equals ten million.

Introduction

April 1999, Bhubaneshwar, Orissa

Bhubaneshwar, capital of the state of Orissa, lies near India's eastern coast on the Bay of Bengal. The center of an ancient kingdom, Bhubaneshwar is said to have once held nearly 7000 temples. Today the jungle has swallowed many of these temples and the ancient kingdom is part of what Indian tourist brochures like to call the country's 'hoary past'. Bhubaneshwar is now a marginal city. The state is poor, the government inefficient, the climate deadly.[1] In a country with more than a third of its 1 billion people living under the poverty line, and a literacy rate of 64 percent, Orissa is considered one of India's least developed regions. In Bhubaneshwar there are not many tourists and there are no McDonalds or Kentucky Fried Chickens. Here, a cold Coca-Cola tastes deliciously exotic. Yet, even in this city, on the periphery of both India and the world, cyberspace is spreading.

Around the corner from the makeshift tribal encampment and the gathering of rickshaw wallahs sleeping in the shade, past the vegetable cart and the tea stall selling samosas and chai for one rupee each, you come across an 'STD booth', a kiosk that retails in local and long-distance calls. Tucked away in the back room is a brand new computer, most likely bought on the gray market – the quasi-underground trade in digital technology. The computer is not yet hooked up to the Internet, and the employees are only now learning how to use even the most basic applications. Yet this does not matter much, since here, as elsewhere in India, info-tech is immediately productive, plugging directly into a micro-commercial culture in which computer time is directly bought and sold. Further down the road, past the temple with its intricate carvings and manicured lawns where old men come to sit and chat, past the paan shops, the

1

occasional goat, the fruit sellers, and the vegetable stalls, the stereo-typical vision of a South Asian street scene is interrupted by a painted sign advertising one of the myriad small courses and companies specializing in software programming, network solutions and multimedia design. The sign bears one of those strange names that now populate the Indian landscape. CYBERCOM…BHARATDATATECH…INFOTRON… the semiotics of cyberpunk incarnated through a creative mixture of Indianized English and IT.

Across the street, in front of the dilapidated theater showing the latest Bollywood production, the newsstands are filled with glossy computer magazines and the papers' matrimonial ads prominently highlight computer education as one of the most desirable traits in a future spouse.

The 'Gayatri Marriage Bureau', an office on the main street leading into town, further supports this mingling of IT and matrimony. Despite its bureaucratic name this is the workplace of an occultist or – as is written on his business card – a consultant for Astrology, Gems, Tantra and Vastu. The 'bureau's' customary activities, offering advice, telling fortunes, reading birth-charts, are now augmented by astrological software and color printouts. To supplement regular business the office also offers access to the Internet and e-mail.

More reliable connections to the web are provided by the various cybercafes sprouting up throughout the city. These range from cramped market shops to plush rooms with new computers and Ikea-style desk chairs, where one can surf the net in air-conditioned comfort, occasionally glancing through the window at the cows wandering listlessly outside.

In Bhubaneshwar, digital technology is integrating almost impercept-ibly into the indigenous culture. Unlike the highly disruptive factories and transportation networks of the industrial age, cyberspace evolves through what might be termed a process of 'soft industrialization'. It spreads unobtrusively throughout the side streets and neighborhoods, flourishing amidst the chaos of local markets, and only surreptitiously alters the preexisting channels of communication and trade.

Despite the lack of upheavals, India is in the midst of a revolution; one that many argue is no less profound than the revolution of 1947 in which India was granted its independence. As Gurcharan Das writes in his book, *India Unbound*, there is 'a soft drama taking place quietly and profoundly in the heart of Indian society. It unfolds every day, in small increments barely visible to the naked eye.'[2] For Das this 'soft drama' has created a situation in which, 'India enters the twenty-first century on the brink of the biggest transformation in its history. The changes are more fundamental than anything that the country has seen, and they hold the potential to transform it into an innovative, energetic economy of the twenty-first century.'[3]

Reporter Cheryl Bentsen, who writes in a CIO magazine field report on the new economy, also called *India Unbound*, shares this same sentiment: 'Traveling through India, I hear one phrase repeated many times – from writers, software engineers, movie producers, investment bankers, members of parliament, shopkeepers and even one crisply tailored hotel driver, who, steering his taxi through streams of mopeds, auto-rickshaws, wobbly trucks, cars, cows and panting dogs in downtown Bangalore, glances in the rearview mirror and announces with pride, "It's our time."'[4]

This new-found confidence and increased aspiration is intimately connected to the promise of digital technology, which many hope will allow the country to leapfrog past the industrial revolution and become, in Bill Gates' words, 'a software superpower'. As Dewang Mehta, former president of the National Association of Software and Service Companies (Nasscom), writes, 'India has a new mantra – Information Technology – and almost everyone has started chanting it.'[5] Atal Bihari Vajpayee, the BJP leader and India's former prime minister, has transformed this mantra into a clever slogan. 'IT', he confidently declares, 'is India's Tomorrow.'[6]

This enthusiastic embrace of high-technology challenges typical conceptions of India, which tend either to pity the country for its underdevelopment and poverty or exalt it for its spiritual superiority. The image of a high-tech whiz has become the icon of a new India, one that defies the stereotypes of snake charmers and starving children. Indeed, the Indian 'techie' has begun to take its place alongside blue jeans, fast food and MTV as one of the key symbols of contemporary capitalism.

With a billion people, the world's largest democracy, an enormous pool of English-speaking engineers, an expanding middle class and one of the greatest untapped markets on the planet, India's encounter with cyberspace has a far-reaching impact on the future of globalization. Yet, how are we to understand this new India and its place within the information age?

According to one quite traditional and still widely held perspective, India's involvement in the IT revolution is best understood as belonging to a process of Westernization, in which an ancient civilization gives way to a culture and way of life that has already been determined elsewhere, by Europe and America. The Indian software whiz – icon of the new – is seen merely as a symbol of the country's growing absorption into a homogenous global monoculture, once identified with the bourgeoisie of Europe, but now equated instead with the pop culture of America.

This notion that India's development is based on the 'Westernization of the world' depends, as will be shown in Chapter 1, on theories which stress the unequal relations between the economic core and its peripheral zones. Due to its position on the margins of the global system, India is seen as mimicking – desperately attempting to catch up with – the more advanced cultures of the West. Implicitly underpinning this belief are three 'core postulates': that globalization radiates out from the core, that the periphery is backward relative to the core, and that what occurs at the edges is only ever of marginal importance to the homogenizing forces of globalization that are being directed by the West.[7]

For anyone visiting India in the mid 1980s this notion of a marginal, backward periphery, trapped in a futile struggle to catch up with the core, might have been confirmed after the first attempt to make a long distance telephone call. Having waited half the day at a crowded telephone exchange for the chance to make an exorbitantly costly, barely audible phone-call, one could hardly help but conclude that India as a modern, technological power was – decades after independence – terribly and hopelessly 'behind'.

In India's protectionist phase (the topic of this book's first two chapters) the telephone system was scandalously poor. For most of its history the Indian government thought of telephones as luxury items and telecommunication infrastructure was given little priority. This neglect was further compounded by the fact that those telephone services that did exist were firmly in the hands of large state monopolies, which were notoriously bureaucratic, inefficient and corrupt.

Discussing this topic in his book *Conversation with Indian Economists*, V.N. Balasubramanyam writes: C.M. Stephen, Prime Minister Indira Gandhi's Minister of Communications, is reported to have 'seethed with indignation when questioned in Parliament about the inadequacies of the telephone system. Telephones were a luxury, not a right, according to the Minister, and any Indian who was not satisfied with the telephone service could return his phone.'[8]

The results of this mindset were all too predictable. 'Until a few decades ago, the telephone service in India was one of the worst in the world.'[9] By 1980 there were only 2.5 million telephones in India, with a mere 12,000 public phones for 700 million people. Even as late as the year 2000, India had less than 30 million phones, which amounted to just over 2 telephones for every 100 people.[10] This should be contrasted with the situation in China which has a similarly sized population but more than 200 million phone lines (the comparison between these two giant neighbors is further explored in Chapter 3).

Yet anyone returning to India in the late 1990s would have found the situation entirely different. Mostly this was due to the country's new policies of liberalization and opening up, whose background and implications are described in Chapter 2. One of the most striking transformations was in the area of technological change. In a sudden burst of creative entrepreneurial innovation – examples of which will appear again and again throughout this book – a far more advanced technological network, cable TV, appeared suddenly, everywhere throughout the country, as if all at once.

In the case of cable TV, the sluggish apathy of state bureaucracy was replaced by the efficiency and speed of private entrepreneurs. The outcome was striking. Telephones first came to India over 150 years ago, introduced by the Raj in 1851. Cable television, on the other hand, is little over a decade old. Yet shockingly, by the turn of the millennium, more people had access to cable TV than to a fixed telephone landline.

The market for cable TV first arose in India in 1991, the year fundamental economic reforms were initiated. The motivating spark was the first Gulf War. Many people, concerned for friends and relatives working in the region, crowded around the few satellite TVs in the country, hoping to catch a signal from STAR TV, CNN or the BBC.

In no time, countless entrepreneurs sensed the opportunity and began an anarchic race to wire their local communities. Though prohibited from broadcasting anything from inside India except for Doordarshan (the state-run station), scores of small-time cable operators – or cable wallahs – found ways to bypass this obstacle, producing videos in India, for example, and sending them to other destinations in Asia, like Hong Kong and Singapore, where they could be transmitted by satellite. Alternatively, operators would simply rent foreign films and television serials, or buy pirated copies, and broadcast them locally.[11]

Cable TV thus proliferated throughout the country in a completely unregulated fashion. As *The Economist* magazine reported in an article entitled *The Wiring of India*, 'the land of the "license raj" somehow forgot to regulate cable'.[12] Kanwal Rekhi – one of Silicon Valley's most famous Indian entrepreneurs (whose massive contributions to the global IT economy are discussed in Chapters 4 and 8) – explains that 'the entire process [of setting up cable TV] went under the radar. The government wasn't even watching. By the time they became aware there were 50 to 60 million subscribers already.'[13]

Today, Indian TV, along with its films, newspapers and magazines, is among the most dynamic media in the world. Every subscriber has access to dozens of stations in numerous languages with shows ranging from mythological epics to contemporary soaps. On Indian MTV, Bolly-wood hits with their elaborate song and dance routines compete with the much more low-key indie pop videos. News channels are filled with current affairs shows, whose depth of engagement and seriousness of debate is only very rarely matched in the developed societies of the 'core'.

In the West, the development of telephones and TV emerged in a strict linear progression. Each new network has been stacked one on top of the other. In Canada, for example, it is inconceivable to have cable TV in your home and not have a telephone line.

On the periphery, however, new technologies do not follow this plodding route of 'normal' progress. Instead, they appear as if spontan-eously through a process which might be called 'anachronistic rupture'. This occurs both at an individual level, when people who have never had any kind of telephone suddenly gain access to the most advanced telecommunication devices, for instance a wireless mobile phone, and also at the societal level, when whole cultures cease to play 'catch up' and begin to 'leapfrog' past the developed world. This is evidenced by the fact that, as Allen Hammond and Elizabeth Jenkins write in an article entitled *Bottom up Digitally-Enabled Development*, 'the growth rate of mobile phones or host computers hooked up to the net is faster in developing countries than in many rich countries'.[14] Chapter 7 discusses the implications of this 'discontinuous growth' on the issue of the digital divide.

From the semiliterate farmer in India who uses IT to access local information, to the gizmo crazy teenager of Shanghai, technology is infiltrating Asian culture with a speed and intensity that is simply unmatched in the West. Technologies like DVD players, video adver-tising screens or robot pets all appeared first on the periphery, but nowhere is this advance of the edges more obvious than with wireless, the latest technological grid, which is spreading throughout Asia with astonishing speed. In India the explosion of wireless is everywhere apparent. In 1999 it was rare for anyone outside the technological elite to own a cell phone. Today, cell phones – at least in the urban centers – are almost ubiquitous. They have become a near necessity in a country where the ability to mobilize business and personal networks is crucial for getting anything done.

Since 1997, the Indian market for cell phones has been growing at 145 percent a year and many predict that this pace will only intensify. Speaking at the 'IT Forum' in Hyderabad, then communications minister Pramod Mahajan predicted that, in 2003 Indian cell phone companies would add 20.5 million new customers; in 2005 the number of phones in the country would exceed 100 million; and by 2020 India would have 500 million mobile phones.[15] While a certain amount of skepticism is always healthy in the face of such government statements, wireless has a host of advantages that lends credibility to these predictions.

By 2003, intense competition in India's private sector led to some of the most spectacular price cuts in the world, with the Reliance group offering mobile telecom rates in India that are among the cheapest anywhere on the planet. The increasing affordability of mobile communication means that mobile phones have the potential to grow exponentially, not only among the enormous emerging market of the middle class, but also among millions of the rural poor, many of whom do not even have access to land lines. This is made more likely by the fact that, as techno-theorist Osama Manzar points out, 'mobile phones do not face a literacy barrier and can easily tap into the intensely oral culture of village life'.[16]

Management consultant and poverty advocate, C.K. Prahalad, who will feature prominently in the pages that follow, argues that this anachronistic 'leapfrogging' is vital to India's development. The biggest mistake India can make, he warns, is to assume that the future emerges as the inevitable next step in linear, historical time. Be very careful, he cautions, not to extrapolate from where you are. To do so is to remain unnecessarily trapped by one's current surroundings, caught within the illusory 'realm of the possible'.[17] In order to create the future, India must 'escape the past'.[18]

This requires that the country start with its aspirations – however 'impossible' they might seem. Those who are constrained by what they deem possible always miss the real source of change, the unexpected realm on the edges, where all the most revolutionary transformations occur.

Kanwal Rekhi, one of the most successful figures of Indian IT, echoes this theme, 'Who would have thought that a world class IT industry was possible in India 10 years ago? Who would have planned it? I used to laugh at it to tell the truth.'[19] Chandrababu Naidu, the tech-savvy ex-minister of Andhra Pradesh, concurs. 'Everybody in India thought: We cannot succeed in IT. We are not America; we're not Singapore. Now everybody thinks we can do it.'[20]

India, advises Prahalad, cannot afford to wait for the future to arrive. Rather it should reach toward it and 'fold the future in'. It is this positive engagement with the future that makes the developing culture of India so exciting and – from a Western perspective – strangely optimistic.

Developed societies, in comparison, tend to be permeated by a sense of already having arrived, which gives rise to the depressing cynicism so noticeable within the cultures of the West. Compared to weary core societies, the periphery has much less invested in the past and is far less shackled to the historical legacy incarnated in technological systems. 'One of the troubles in the US with the wireless industry', says writer and internet consultant Madanmohan Rao, 'is that people are already happy with all that copper – they've got a good land connection, what do they need cell phones for?'[21]

During 11–14 February 2003, Nasscom convened its annual industry-wide conference on 'Leadership convention' at the Hotel Oberoi in Mumbai. Parallel sessions were held on topics such as 'Indian IT as a global brand' and 'global IT trends'. There were over half a dozen 'country forums' featuring speakers from Asia, Europe and North America. The atmosphere and tone heralded an increasing professionalism and a kind of comfortable settling-in of India's IT industry, whose history is described in Chapter 5.

On the hotel's top floor, where a large window looked out onto the Arabian sea, a relatively small room was devoted to sessions on issues and trends raised by the newly emerging ITES-BPO sector, a business area, which is more familiar to Western audiences as 'offshore outsourcing' (discussed in detail in Chapter 6). Here the calm professionalism that permeated the rest of the Nasscom conference gave way to an electrifying buzz. Sessions that featured partners at McKinsey and the president of GE had standing room only. Audience members had to squeeze into the packed room, teetering on tiptoes to hear the presentations.

Among the topics being discussed were strategies for countering the protectionist backlash expected from the West. From an outsider's perspective, at the time, these fears seemed unnecessarily defensive – even paranoid.

A year later, however, magazines, newspaper articles, blogs and TV debates all seemed obsessed with 'outsourcing to India'. From being pitied for its marginality and backwardness, India had suddenly become an object of growing alarm. The country was now being accused of 'stealing' jobs from the most advanced sectors of the most advanced economies in the world. In America, the topic of outsourcing and its effect on local jobs proved to be among the most important issues of the 2004 presidential election.

In an attempt to curb this frenzied reaction against free trade, Indian-American economist Jagdish Bhagwati, in a lead editorial in the *New York Times*, entitled *Why Your Job isn't Moving to Bangalore* argues that it is not India but technological change itself that is responsible for shifting the job market in America's dynamic economy.[22] Yet, this technology-driven transformation has been guided by India, whose own IT industry has been able to anticipate the future and thus place itself – as the increasing hype testifies – at the cutting edge of the global economy. As we will see in the final chapter of this book, it is not for the first time that the globalization of the West required the reluctant acceptance of innovations that have emerged out of India. For centuries, India has contributed to a creative 'hollowing-out' of the core which, although resisted at every stage, has produced our dynamic and diverse globalization. The future always proceeds from the edges.

1
The Idea of Westernization

The bourgeoisie, by the rapid improvement of all instruments of production, by the immensely facilitated means of communication, draws all, even the most barbarian nations into civilization. The cheap prices of its commodities are the heavy artillery with which it batters down all Chinese walls...It compels all nations, on pain of extinction, to adopt the bourgeois mode of production; it compels them to introduce what it calls civilization into their midst, i.e. to become bourgeois themselves. In one word, it creates a world after its own image...Just as it has made the country dependent on the towns, so it has made barbarian and semi barbarian countries dependent on the civilized ones, nations of peasants on nations of bourgeois, the East on the West.

– Karl Marx and Frederick Engels,
The Communist Manifesto

Glottopolitics

Voices – an NGO that advocates the use of media as a vehicle of empowerment and an agent for social change – has been working since the turn of the millennium, in partnership with UNESCO and another NGO called 'Myrada', on a pilot project in a poor village in India, located about three hours from Bangalore. The project aims to aid local development by using audio production as a platform for community-based media. It employs a wide spectrum of technologies, from the Internet to megaphones, and involves everything from training people to use computers, to setting up an audio production center designed to serve the village. Participants, organized primarily by local 'self-help groups', are given the technical skills to create local content. They are encouraged, for example, to surf the web looking for valuable material and then to repackage that information into an audio format that can be made available to the wider community. Due to India's licensing regulations, the community is not allowed to broadcast its own radio programs, so local producers employ a variety of low-tech solutions including audiocassettes, loud speakers and cablecast to reach their target audience. In this way, the village is able to gain access to material which they themselves deem useful to their existence. Examples include the weather forecast, information on government schemes, knowledge of healthcare issues and strategies for dealing with drought.

The activists who initiated the project assumed that in this poor community, composed mostly of coolies or agriculture laborers, one of the most basic needs would be for local language media. Many in the community are either illiterate or semiliterate and do not speak, read or write in either Hindi or English – India's two official tongues. The NGOs believed, therefore, that creating media in the villagers' native languages was essential to bridging the digital – and even wider information – divide. They were, therefore, somewhat surprised to find many people in the community expressing their desire for English. 'We were pushing the need for local language technologies,' says Ashish Sen of Voices, 'but they [the villagers] have very strongly articulated the need to be fluent in English.'[1] What Voices discovered is that, even after a half century of independence, there is an awareness among India's rural poor, that to increase one's prospects and connect with the outside world requires knowledge of English – the world's global tongue.

In India, Sanskrit – the language of the Vedas and of the sacred mantras – is renowned for its mystic power. The power of English, on the other hand, is a worldly one. India is a multilingual society in which different languages serve different functions. This fact, which is reflected in the official 'three-language policy', has created a pyramid structure in the linguistic texture of Indian life. In this pyramid, the regional languages, which are used in the home, in day-to-day living, and at the lower levels of education, occupy the expansive lower tier. Hindi, the language of New Delhi – the seat of political power – has a place somewhere in the middle. At the apex of the pyramid is English. 'If you weigh languages in terms of powers they wield,' writes author Khushwant Singh, 'you will see that English overweighs all the other Indian languages...put together.'[2]

In contemporary India, 'all higher order activities in the domains of education, commerce, and administration have to be negotiated and performed in English'.[3] Higher degrees, especially in social sciences and even more importantly in science and technology, are available only in English. 'One cannot become a doctor, engineer, lawyer, bureaucrat, scientist, pilot or the CEO of a corporate house unless one has a proven proficiency in English.'[4]

For these reasons the English language has become intimately connected to development, growth, globalization, prosperity and prestige. 'Those who knew English became a symbol of modernization', writes Braj Kachru, a scholar of world English. 'English internationalizes one's outlook...it permits one to open the linguistic gates to international business, technology, science and travel.'[5] With these associations it should come as no surprise that, as Kachru writes, 'in many respects the roots of English are deeper now than they were during the colonial period'.[6]

From at least as early as the 18th century, world trade has been intimately connected with the English language. Today it is the lingua franca of business deals and the banking system, both in the English and the non-English speaking world. 'English', writes Barbara Wallraff in an article entitled *What Global Language*, 'is the working language of the Asian trade group ASEAN – the official language of the European Central Bank, even though the bank is in Frankfurt and neither Britain nor any other predominantly English-speaking country is a member of the European Monetary Union.'[7]

Even more important than its links with global trade is the fact that English has long been considered 'the language of technology from Silicon Valley to Shanghai'.[8] When Thomas Edison devised the phonograph, his first words – 'what God hath wrought' – were English words. The first radio broadcast was in English, as was the first telephone call. Five of the largest television companies in the world (CBS, NBC, ABC, BBC, CBC) broadcast their programs in English. CNN beams out the world's news – in English. The two largest wire services (Reuters and Associated Press) are English, as is three-quarters of the world's mail, and its telexes and cables.[9] Over half of the world's technical and scientific journals are published in English. In the messages sent by Voyager 1, the head of the UN made an address whose intended recipients were distant alien races. His words, which began 'I send greetings on behalf of the people of our planet' were English words. Finally, though it is deemed the most globalized of technologies, a commonly repeated statistic claims that more than 80 percent of the data on the Internet is stored in English.

English is a global language not because a majority of the planet's population speaks English (in fact, certain statistical evidence claims that over 90 percent do not), nor because English is the most widely used language in the world (that honor goes to Mandarin, a language with almost twice as many speakers as English). Rather, English is the language of globalization because it dominates the spheres of greatest influence – commerce, communication, science and technology.[10]

This dominance of English is often taken as evidence that globalization is a process that radiates out from the core, involving the spread of a homogenous monoculture, which seeps out from the West, eventually to envelop the world. Kachru uses the term 'glottopolitics' to describe the way in which English is invariably 'associated with an unprecedented form of linguistic and cultural colonization'.[11] Though glottopolitics was prevalent throughout the British Empire, its contemporary critics generally target the influence of America.

Benjamin Barber's famous book *Jihad versus McWorld* warns against the onrushing economic and technological forces which threaten to unite the planet into one monotonous whole. Critics in this vein, concerned about the increasing homogeneity of global culture, worry that as English – the language of the world's only remaining superpower – is beamed out through TV sitcoms, ads and Hollywood movies it is, to use Thomas de Quincy's words, 'traveling fast towards the fulfillment of its destiny... running forward towards its ultimate mission of eating up, like Aaron's rod, all other languages'.[12] Barber writes:

> English has become the world's primary transnational language in culture, the arts as well as in science, technology, commerce, transportation and banking. Music television sings, shouts, and raps in English. French cinema ads are now frequently in English. New Information Age critics attack the hegemony of CNN and the BBC World Service but they attack it in English. Somalian clan leaders and Haitian attaches curse America, for the benefit of the media, in English.[13]

For Barber, then, English is both the language of the dominant culture and also – more ominously – the language of cultural opposition and dissent. Used among international businessmen as well as angry teenagers, English is the language of an emerging, monocultural 'McWorld'.

Modern capitalism and the Westernization of the world

This contemporary fear, that the planet is being swallowed by an increasingly homogeneous global culture that radiates out from the core, stems from the older and even more entrenched notion that globalization involves the 'Westernization of the world'. This idea has become so familiar and widespread that many accept it as an unquestioned truth. Today it is common for people to label all signs and expressions of modernity – from the skyscrapers of Shanghai to the futuristic neon-laden streets of Japan and Hong Kong, to the techno parks of India – 'Western', despite the fact that there is really nothing in the West that compares with the science-fiction landscapes of Asia.

The idea of Westernization has its theoretical roots in the modernist writings of Karl Marx and Max Weber, both of whom argued that the origins of capitalism were intimately linked to the history and culture of Europe and America. For these theorists of modern capitalism, it is not contingent but necessary that the capitalist system arose in the West because it is only Western culture that possesses the internal dynamics necessary for capitalism to take hold. Capitalism, according to both Marx and Weber, is the direct and inevitable product of the history of the Western world. To quote World-Systems analyst Samir Amin, 'the social theory produced by capitalism gradually reached the conclusion that the history of Europe was exceptional, not in the sense that the modern world (that is to say, capitalism) was constituted there, which is in itself an undeniable fact, but because it could not have been born elsewhere'.[14]

For these modern theorists, technological progress and capitalist economics could never arise indigenously in a place like India. Rather, if India were to become modern, it would have to be 'Westernized'. In the East, capitalism spread by subsuming the traditional cultures it happened to encounter. It dominated, both economically and culturally, as an invader from outside, imposing its foreign values and alien way of life.

Few theorists have been as influential in tying Western culture to the capitalist economy as Max Weber. For Weber the question 'why did capitalism originate in the cities of the Occident?' was answered by the insight that 'the city of the Occident is unique among all other cities of the world... [in that] it has been the major theatre of Christianity'.[15] Capitalism belonged to the West, according to Weber, because its 'spirit' corresponds to the beliefs, practices and ethos of a particular strand in the Christian tradition. Christianity – or more specifically Protestantism – was the first religion to have developed an economic ethic conducive to the capitalist way of life. Weber called this the Protestant Ethic and in his famous work *The Protestant Ethic and the Spirit of Capitalism* wrote that this 'religious determination of life-conduct' constituted the very 'spirit of capitalism'.[16]

Weber's work was influential in defining not only the modern economy but also the culture of the West. The Protestant Ethic depends, first of all, according to Weber, on the 'disenchantment of the world'.[17] Christianity, like its precursor Judaism, is founded on a grand exorcism, which banishes magic, expels spirits and strips objects of their animistic power. Outlawing the sorcerer, it puts the prophet in charge of a world that is 'disenchanted of its gods and demons'.[18] The sacred, now captured by the monotheistic rule of transcendence, retreats from the everyday. In place of a magical world, there arises a rational, intellectualized cosmos, which is, for the first time, capable of being measured, quantified and governed according to a transcendent law.

Seeking to distance itself from mysticism as well as from magic, the Judeo-Christian tradition shuns not only sorcery but also what Weber calls the 'exemplary prophet', a figure that is characterized by an ecstatic relation to the divine.[19] Trance and possession are criticized since, in these states the individual attempts to flee from the world that God has created. What is required instead is the 'emissary prophet' who no longer functions as God's 'vessel', but has become instead, his 'tool'.[20] With careful sobriety the emissary prophet encounters the divine as the giver of the law. Working as God's instrument he teaches his population to mold life in accordance with the divine will.

The result is the ethos of 'active – or worldly – asceticism', Weber's name for an asceticism that has turned away from a contemplative 'flight from the world' and dedicated itself instead to 'work in the world'.[21] Placed in the role of custodian or guardian, the active ascetic seeks to 'create the kingdom of heaven on earth', transforming the world through the activity of labor. With the active ascetic, then, economic activity ceases to be a hindrance to religious life and becomes instead a sacred duty.[22]

Eastern religions differ fundamentally from this ethos of the active ascetic since, according to Weber, they tend to treat the 'everyday world' with ambivalence – even contempt. Rather than seeking to change the world, Eastern culture aims to flee from it. This aspiration for a mystical 'other-worldly escape' creates a belief in an immutable world order, a desire to flee the mundane and a positive lack of interest in economic life. For this reason, Weber believed, the cultures of the East create static societies that do not have the dynamism necessary for the capitalist system to take hold. Capitalism, which embodies not only an economic ethic but also a religious attitude, has no choice but to bring its own culture – that is to say Western, Christian culture – with it as it spreads.

The notion of Westernization is also at the heart of the dialectical theory of history put forth by Karl Marx. As is well known, Marxian historical materialism is based on the successive development of different organizations or 'modes of production'.[23] In the Preface to *Contribution to a Critique of Political Economy*, Marx divides Western history into three such phases or modes: the ancient, the feudal and the bourgeois. These three phases are related both historically and dialectically for Marx, such that each successive mode develops out of the internal contradictions of the last. 'New superior relations of production', he writes, 'never replace older ones before the material conditions of their existence have matured within the framework of the old society.'[24] Joseph Schumpeter summarizes the point as follows: 'The forms of production themselves have a logic of their own; that is to say, they change according to necessities inherent in them so as to produce their own workings.'[25] According to Marxism, then, the ancient necessarily gave way to the feudal, and the feudal necessarily gave way to the bourgeois. Capitalism is thus an inevitable product of the dialectical unfolding of history. It 'is itself', writes Marx, 'the product of a long course of development, of a series of revolutions in the modes of production and exchange'.[26]

This 'series of revolutions' is, for Marx, unique to the West. The dialectical sequence does not arise elsewhere unless it is imposed from outside. The East, Marx argues, is a static society. In his writings it is characterized by the 'Asiatic mode of production', a social system that is entirely different from the other social modes. Whereas the Western modes of production are related dialectically and therefore historically, the Asiatic mode, writes Marx, is determined by geography alone.

In the case of India, the Asiatic mode of production is explained through an analysis of what Marx saw as the dominant structure organizing production, the self-sufficient communal village. Marx's analysis of these villages focuses on three main interlocking themes. First, in these villages labor is divided through a highly coded structure based on caste. The rigidity of this codification results in a situation in which 'the law that regulates the division of labor in the community acts with the irresistible authority of a law of Nature'.[27] It thus produces 'an unalterable division of labor, which serves, whenever a new community is started, as a plan and scheme ready cut and dried'.[28] Second, the self-sufficiency of these villages means that most of what is produced is 'destined for direct use by the community itself, and does not take the form of a commodity'.[29] In these villages it is only the surplus that becomes a commodity and this is paid directly to the state in the form of rent. Finally, the Asiatic mode rests on the absence of private property in land. 'This', Marx writes in a letter to Engels, 'is the real key even to the

Oriental heaven'[30] since without private property the capitalist system simply cannot arise.

These three factors ensure that the Indian village can maintain itself in a state of constant equilibrium. Lacking the internal struggles of class conflict, the villages continually reproduce themselves according to the same structure. For Marx 'this simplicity supplies the secret of the unchangeableness' of the Asiatic mode. In the East, the material processes that form the 'ground of history' remain unaltered regardless of any political turmoil that may occur at the level of the state. 'The structure of the economic elements of society', writes Marx, 'remains untouched by the storm-clouds of the political sky.'[31]

The fact that political changes do nothing to transform the internal dynamics of the economic structure on the 'ground' means that for Marx, the Asiatic mode of production falls outside 'earthly' history. Lacking any change at the production level, Marx concludes that societies based on the Asiatic mode are unchanging, nondialectical and exterior to historical time.[32] 'Indian society has no history at all,' he writes, 'at least no known history. What we call its history is but the history of successive invaders who founded their empires on the passive basis of that unresisting and unchanging society.'[33]

Though inherently tied to the history and culture of the West, theorists of modern capitalism believed that this Western social system had, from its inception, global aspirations. Marx, in particular, maintained that the need for global expansion was an immanent feature of bourgeois society. 'The need of a constantly expanding market for its products chases the bourgeoisie over the whole surface of the globe. It must nestle everywhere, settle everywhere, establish connections everywhere.'[34]

This image of capitalism has the traits of a gothic nightmare. It appears in Marx's writings as an unstoppable entity called up by spells from the 'nether world' that have escaped the 'sorcerer's control'.[35] Growing ever outward, it subsumes the world, 'sucking in living labor as its soul, vampire-like'[36] as it spreads. Capitalism, according to this vision, is inherently destined to reformat the planet 'in its own image'. This image is that of the West.

Protecting the periphery

Outside the West, on the periphery, the notion of Westernization is bound to the idea of development, which in turn is bound to the 'idea' of the West. The intense linkage between these concepts was already well established by the late 19th century. By the time Marx was writing,

the Western – and in particular the British – experience of industrialization was believed to be the model that all societies hoping for economic
growth and technological progress would have to follow. This pattern
of historical progress, it was widely believed, followed a linear 'arrow of
time', which gradually unfolded through a series of steps or stages. This
pattern had already manifested itself in Europe through the vast changes
unleashed by the industrial revolution, in which feudal, agriculture
societies were transformed through manufacturing and factory-based
production. Development was a game of catch up. No matter where, it
could only occur by imitating this Western model. As Marx wrote in the
preface to *Capital* Volume 1: 'The country that is more developed indus-
trially only shows, to the less developed, the image of its own future.'[37]

W.W. Rostow's book *The Stages of Economic Growth: A Noncommunist
Manifesto*, which was published in 1960, contained one of the most
influential theories of this linear model of development. Rostow argued
that all societies will eventually pass through a series of developmental
stages which imitate – or at least conform to – a Western model. Rostow's
book outlined the five stages of 'economic development all societies
eventually experience as they mature into industrialized developed
countries: tradition, the preconditions for takeoff, the takeoff, the drive
to maturity, and the age of mass consumption'.[38] Rostow based these
five stages on the experience of Great Britain which was, according to
Immanuel Wallerstein, 'the crucial example since it was defined as being
the first state to embark on the evolutionary path of the modern industrial world. The inference, quite overtly drawn, was that this path was a
model, to be copied by other states.'[39]

In the period following World War II, however, a new theory emerged
which would fundamentally challenge this notion of a linear pattern of
economic growth. It called itself 'World Systems theory' and, unlike the
Marxist doctrine out of which it grew, it focused on the geography of
the economy, rather than its history. World Systems theory challenged
the accepted notion of development by raising two major objections.

First, it rejected any analysis that was based on the nation-state, arguing
instead that capitalism was a 'world system from the start'. For World
Systems theory the emergence of capitalism was necessarily coterminous
with European mastery of the ocean and the onset of colonialism. The
'crystallization of capitalist society in Europe and the European conquest
of the world', wrote Samir Amin, one of the theory's more vocal proponents, 'are two dimensions of the same development, and theories that
separate them in order to privilege one over the other are not only
insufficient and distorting but also frankly unscientific'.[40] Capitalism,

according to World Systems theory, could never have developed within the confines of a domestic market. From its very establishment it spilled over state borders encompassing indigenous cultures and heterogeneous social structures into a single global economy. As Immanuel Wallerstein, probably the leading World Systems theorist, writes, 'capitalism was from the beginning an affair of the world economy and not of Nation States'.[41]

Secondly, World Systems theory refused to see capitalism as a particular stage of development. Pointing to the historical simultaneity between developed and underdeveloped regions, it forcefully maintained that inequality was a necessary and inevitable part of the capitalist system itself. 'Underdevelopment' was not simply a temporal stage that would be remedied by progress but rather a constant feature of the world economy. 'Economic backwardness', the theory contended, 'was not at all a matter of starting late but was instead itself a condition produced in the course of and as a result of the rise of capitalism.'[42]

For World Systems theory it is thus no accident that the countries of the 'third world' have failed to 'catch up' to the West. The very structure of capitalism destines these countries to an inevitable lack of prosperity and a never ending position of economic dependence.[43] The idea that the third-world countries 'will progress to the extent that they imitate the West'[44] is a misguided and dangerous myth, rooted in the refusal to recognize the fundamental polarity that forms the ground of the modern world system – the distinction between core and periphery zones. The growth of global capitalism, writes Amin, 'progressively created a growing polarization at the heart of the system, crystallizing the capitalist world into fully developed centers and peripheries incapable of closing the ever widening gap, making this contradiction within "actually existing" capitalism – a contradiction insurmountable within the framework of the capitalist system – the major and most explosive contradiction of our time'.[45]

It did not take long for these ideas to take hold in the underdeveloped periphery. Many countries in Africa, Latin America and Asia were frustrated by their position in the world and suspected that their own lack of growth was somehow connected to the Western dominance of the global market. This suspicion was consolidated by what became known as 'Dependency theory'; an offshoot of World Systems analysis that began in Latin America, and by the 1960s, was sweeping the third world. Dependency theory argued that underdevelopment was a direct result of the structural inequalities that existed between exporters of raw materials and exporters of manufactured goods. The injustices of global trade were such that the prices of goods manufactured in the developed world would continue to rise while the value of the agricultural products and raw materials produced in the third world would continue to decline. Because of these 'unequal terms of trade', 'poor countries were permanently imprisoned in a state of dependency on the rich'.[46]

In both World systems and Dependency theory, the historical notion of stages of growth did not so much disappear as become spatially transcoded. It is through this spatial transcoding that the contemporary idea of the global periphery emerged. When the economy was looked at from a global perspective, these theorists argued, one could see each stage occurring simultaneously in different geographic regions. Far from being accidentally produced, this difference was an integral part of the capitalist system. Capitalism, which functions inherently with an international division of labor, divided the world into core and periphery zones that were structurally and systemically interrelated. 'Without peripheries, no cores,' says World Systems theory, 'without both, no capitalist development.'[47]

In India this idea seemed particularly persuasive since it reinforced widely accepted assumptions about the workings of the British Empire. According to these, the Empire operated by creating an international division of labor in which manufactured goods were produced and exported by the core while the colonies – or 'peripheral zones' – supplied the core with the resources it needed to maintain this industrial base. Though in the post-war period the West was forced to withdraw politically from countries in Asia and Africa, it was able to maintain this structure of trade and thereby – many suspected – preserve its wealth. The more 'advanced' countries continued to benefit by exporting manufactured goods while counting on the underdeveloped to import these goods whilst exporting raw materials. Thus the capitalist system – even in the postcolonial period – continued to favor the Western core at the expense of the regions on the edges.

The periphery, understood as that which occurs on the sidelines, margins or edges, is generally taken to be a spatial designation, though its precise location in the realm of geopolitics is unclear. Though associated more with the East than the West, the South than the North, peripheral zones are unstable, constantly shifting and capable of appearing anywhere. It is common now to speak, for example, of America's 'internal' peripheries.

Such dichotomies arise from the fact that although the notion of 'the periphery' is a spatial designation, it remains a prisoner of its temporal connotations. The idea of the global periphery arose as an attempt to challenge accepted ideas of historical progress. Yet, it nevertheless retains an inherent vision of development, which implies – and reinforces – a particular conception of continuous, progressive time.

Even when deterritorialized – abstracted from any particular place – 'the periphery' still suggested that certain geographical regions were locked in the past, their development retarded so that they remained caught 'behind' the core. With their growth inhibited, even if by a contemporaneous global structure, peripheral regions were still considered embodiments of a more 'primitive' version of 'advanced' countries primarily located in Europe and North America.

The newly independent countries on the periphery saw themselves trapped in a desperate situation. Development – still conceived as catching up with the West – required that they mimic more 'developed' countries in their path of industrialization, moving gradually, step by step, from exporting raw materials to industrial manufacturing and finally services. At the same time, however, Dependency theory told them that the economic imperialism of the global capitalist system made it impossible for them to progress through this sequence.

The perceived solution was to close the door on this structural inequality and 'protect' the country from the outside. The recommended approach was to severely limit foreign investment in favor of large-scale state investment and to adopt policies of import substitution, with foreign imports replaced by locally produced goods. The goal was to create closed economies that were able to function autonomously, cut off from the injustices of global trade. The periphery scrambled to erect barriers that would shelter them from the victimization imposed by world markets.

In India these ideas were embraced wholeheartedly by the country's founders. One of the strongest components of Gandhi's 'Quit India' movement was the ideal of *swadeshi* (or self-reliance). Gandhi believed that economic freedom would come from the self-sufficiency of India's myriad villages. Advocating 'simple home production of basic goods, self-sufficiency in the village, and a spinning wheel in every hut,'[48] he propagated 'a vision of a republic of self sufficient villages employing indigenous technologies, materials and talents'.[49] His romantic attachment to the simplicity of the past led him to promote 'hand-spinning, hand weaving, hand-pounding of rice, hand grinding of corn and traditional oil pressing...The traditional old implements, the plough and the spinning wheel, have made our wisdom and welfare', he said. 'We must gradually return to the old simplicity...The railways, telegraphs, hospitals, lawyers, doctors – all have to go...'[50]

In his personal life, Gandhi expressed his commitment to this ideal through an almost fanatical devotion to *khadi* (spun cloth). For Gandhi nothing made India's economic enslavement more clear than the fact that a country so rich in textiles was importing cloth from the mills of Manchester which were driving England's own industrialization. 'Why should colonial India export cotton to Manchester, only to import it back in the form of expensive clothing?'[51] he asked. Gandhi encouraged his followers to spend time every day spinning and insisted on wearing only simple garments made of spun cloth – even when touring the West. The fundamental importance of this symbol of swadeshi is clear from the fact that the image of the spinning wheel was prominently displayed on the Indian national flag.

Nehru argued against Gandhi's Indian version of Luddism, but he shared Gandhi's aspirations for a self-reliant economy. For Nehru, however, the dream of self-sufficiency would not be achieved through the 'backwardness' of the local village. Instead, Nehru created a program for rapid industrialization that would be achieved through the guiding hand of the state.

In this, Nehru was influenced by the idea, which surfaced in the period immediately following the war, that state interventionism could lead to shortcuts, enabling countries to speed up the development process by 'leapfrogging' past certain stages of growth. This theory emerged initially with reference to the Axis countries that were destroyed in the war. In an influential essay on the subject entitled *Economic Backwardness in Historical Perspective* (1951), Alexander Gerschenkron – who was one of the principal advocates of the linear stage theory of development – wrote that 'latecomers' such as Germany, Russia, France and Japan caught at a less advanced stage could speed up industrialization through large-scale intervention by the state. By the late 1950s the idea was picked up by the newly emerging field of development economics, which argued that the principle of 'leapfrogging' through state intervention applied not only to latecomers but also – and even more urgently – to 'late-late comers'[52] such as India.

The notion of a state-led model of industrial growth was attractive to Nehru since it married his desire for technological progress with his postcolonial suspicion of Western imperialism and his left wing distaste for the free market and the open flows of global trade. Though Nehru was friendly with many of the country's leading industrial families, he had an avid disdain for the world of business. 'Never talk to me about profit' he is quoted as saying 'it is a dirty word'.[53]

Nehru's suspicion of profit, the private sector and global trade was fed from many streams: Gandhi's influence, caste prejudice, his Harvard and Cambridge education, and his encounters with the Fabians. Gurcharan

Das describes the resulting amalgam as an 'intellectual joint venture between Keynesian macroeconomics, Stalinist public investment policy, and Gandhian rural development'.[54] The outcome was a Leninist-type economy, in which – though the market was allowed to function in part – the government controlled the economy's 'commanding heights'.[55]

In the period after the war, Nehru's ideas were common – at times scarcely contested – throughout the developing world. What makes India somewhat unique, however, is the purity and longevity of the theories' implementation. When countries in Asia and Latin America were already beginning to open their doors, India closed hers even tighter. The Nehruvian system, which emphasized distribution of wealth over wealth creation and planning over the market, was further entrenched by his daughter Indira. 'Indira Gandhi's government became even more rigid, introduced more controls, and became bureaucratic and authoritarian. It nationalized banks, discouraged foreign investment, and placed more hurdles before domestic enterprise.'[56] Indeed, it is Indira's refusal to change course, rather than Nehru's initial policy position, that is more regularly blamed for India's long history of economic mismanagement. 'Modern India's tragedy is not that we adopted the wrong economic model in the 1950's, but that we did not reverse direction after 1965.'[57]

In the late 1960s, 70s and 80s India withdrew even further from world trade. Yet, it was not only the Nehru family that was to blame. When Indira Gandhi was booted out of power after the Emergency, the Janata Party took over. During their short rule from 1977–1980 they presided over India's most notorious – and extreme – examples of protectionism. It was at this point that 'foreign firms that refused to share their technology with local champions' were expelled. 'IBM packed its bags, as did Coca Cola – penalized for its refusal to reveal its sacred and jealously guarded formula.'[58] These bans lasted over a decade. As late as 1989, travelers leaving India for Nepal were met at the border crossing by scores of young Nepali boys profiting from this protectionism, by doing a vibrant trade in Coke.

The lack of these giant global brands inside India was a sign of the country's ruthlessness in guarding its culture against the coming McWorld. For the first four decades of its independence, India was a model of an inward looking, closed economy that sought to protect its culture and economy from the Westernization of the world.

2
Opening Up

Self-reliance means trade ...

– Manmohan Singh quoted in
The Commanding Heights

The cyber-coolie debate

November 2003. For the past few months, a lively debate on India's relation to the culture and economics of global capitalism has been brewing in the pages of *The Times Literary Supplement*. The debate was sparked by Susan Sontag's essay entitled *The World as India*. Her focus was a fairly obscure argument about literary translation – hardly a topic one associates with heated, public controversy. Yet, in the midst of the article, Sontag, somewhat surprisingly, given her notorious post-9/11 criticisms of America, confesses her pride in English, a language, she writes, 'seemingly identical now with the world dominance of the colossal and unique superpower of which I am a citizen'.[1] Sontag confesses that her admiration for this 'new international language' is intimate and direct. 'Each day I sit down to write I marvel at the richness of the thousand-year-old language I am privileged to use.'[2]

Though a native speaker herself, Sontag does not see English as a national, ethnic or authentic tongue. Instead, she praises English precisely for its 'inauthenticity'. Like many before her who have delighted in the impurity of English, Sontag points to India as the example to prove her point. In particular, she draws attention to call centers, which she describes as 'a flourishing enterprise in the multi-billion-dollar software industry now so important to the Indian economy'.[3]

Sontag is right to note that India has recently experienced an explosive rise of call centers and back offices, due – in large part at least – to the country's English-language resources. In an article entitled *Outsourcing India, The Economist* magazine explains the appeal: 'First world companies do lots of things that are expensive and necessary, and yet "peripheral" to their core competence. The main requirement for these tasks is an intelligent English speaking workforce – which India has in abundance, at a small fraction of rich country wages.'[4]

The enormous potential for India's English-speaking population, in what Indian business circles call 'IT enabled services', is described by Michael Dertouzos, director of MIT's laboratory of computer science. According to *The Economist*, 'Dertouzos reckons that India has some 50M English speakers who could each earn US$20,000 a year – making a total of US$1 trillion, twice India's current GDP – doing "office work proffered across space and time"'.[5]

These projections are, of course, highly optimistic, and the techno-utopian tone strikes a discordant cord with those who see in the back offices of India only the latest example of an international division of labor, in which the periphery is stuck doing peripheral tasks.

It is precisely this attitude that prompted Harish Trivedi, a critic and academic working in New Delhi, to respond to Sontag's essay. In an 'angry rejoinder'[6], Trivedi accuses Sontag of being 'unconscious to the current neocolonialism of America' and blind to the past colonialism of Britain.[7] He reprimands her enthusiasm for both English and call centers accusing her of promoting a cultural and economic imperialism. Making use of the typical creativity of Indian-English, Trivedi calls the call center workers 'the cyber-coolies of our global age'[8] and portrays them as the mistreated drones of the digital sweatshops. 'They work', he writes, 'not on sugar plantations but on flickering screens, lashed into submission through vigilant and punitive monitoring, each slip in accent or lapse in pretence meaning a cut in wages.'[9]

For Trivedi the 'brutal exploitation' of this economic arrangement is matched and intertwined with India's relationship to the English language. 'The vital question', he writes, is 'should we let English continue to rule over us so that we may remain at the beck and call of the Anglophone West, eager to pick up the crumbs of cheap outsourcing?'[10] Global English, like global capitalism, is producing not so much 'The World as India', according to Trivedi, but rather 'The World as (One Big) America'.[11]

A few weeks later a reply in the letter page by Apratim Barua admonishes Trivedi's for his 'postcolonial fury'.[12] Barua, who was born in independent India, has little sympathy for Trivedi's linguistic nationalism. 'The notion of a single "national" language', he writes, 'is derived from concepts rooted in nineteenth-century European nationalism, when distinct nation-states were conceived along ethnic, linguistic and territorial lines. This form of racist nationalism never did suit India, which has always been a polyglot nation.'[13] Instead of playing the victim, he attributes English's success in India to both the language and the country's increasing cosmopolitanism. 'English is an Indian language like any other,' writes Barua. 'It might have been the language of the conqueror a long time ago, but for those of us born in an independent India, that is all a part of history.'[14]

Neither does Barua buy Trivedi's 'cyber-coolie argument' – though he expresses an admiration for the phrase. The argument, he claims, is based on a contradiction. 'Sweatshop laborers are exploited precisely because an abundance of competition makes them easily replaceable. If this were the case, however, it would undercut Trivedi's claim that "English is an elitist language restricted to a narrow upper-class".'[15]

Trivedi replies with another scolding. Barua, his former student, 'should show some respect for provenance and history'.[16] The idea that the British 'conquest and rule of India' is easily brushed away is, he writes, to treat history as if it 'were a junk email spam that one could delete with one impatient click'.[17] Invoking Rushdie, he claims that Barua is 'self-avowedly one of midnight's children'.[18] Yet he fears that the darkness of his birth has blinkered his vision. The letter ends with the hope that Barua awaken to 'the clear light of post-colonial day'.[19]

This heated – and lyrical – exchange prompts yet another letter from author and businessman Gurcharan Das. Like Barua, Das, who himself has been associated with a number of call centers, sides with Sontag as she celebrates 'the success of Indians in harvesting their legendary English speaking skills in the global economy through call centers and other services'.[20] Dismissing Trivedi's depiction of cyber-coolies as 'truly bizarre', Das appeals to economics, arguing that IT enabled services will 'create an enormous number of jobs in India' while providing for their employees the 'exciting chance to work with the world's top brands and acquire new skills to make a career in the global economy'.[21]

Das' economic argument, however, is also a cultural one. 'At the root of the dispute', he writes, 'is ownership of the English language. Today's confident young Indians view English as a functional skill, not unlike Windows or learning to write an invoice. When they speak English they feel they own it.'[22] By contrast, 'HarishTrivedi's neocolonial English flies the Anglo-American flag'. For Das it is the minds of these 'cyber-coolies' that 'seem to be decolonized, whereas Trivedi's is stuck in a post-colonial past…So, who is the coolie?' he asks. 'Not the confident young person at the call center with her liberated attitude to English, but Harish Trivedi, whose mind remains colonized in the old linguistic categories of post-colonial, pre-reform India.'[23]

This debate encapsulates a clash of ideas that has been influencing India for at least half a century. In its early years, independent India sought to assert itself against what it perceived as the dominance of the West by closing its doors and protecting its own autonomous language and markets from the ever-encroaching forces of imperialism. In both the linguistic and economic spheres, however, this strategy has failed and was, eventually, abandoned. Contemporary Indian cyberculture prospers through what Das calls a 'decolonized mindset', which has opened itself up to the culture – and the markets – of the world.

1600: Empire

There is no doubt that English in India is a legacy of the Raj. 'English came to Massachusetts the same way it did to Mumbai: on a British ship,' says Joshua Fishman in an article entitled *The New Linguistic Order*. 'For all the talk of Microsoft and Disney, the vast reach of English owes its origins to centuries of colonization by England.'[24] According to Braj Kachru, 'it is customary to trace the roots of English in the Indian subcontinent to 31 December 1600, when Queen Elizabeth I granted a charter to a few merchants of the City of London giving them a monopoly on trade with India and the East'.[25] Since trade is impossible without communication, the East India Company required, from its inception, a handful of inter-preters who could translate between the English merchants and their Indian counterparts. The company had no desire other than to fulfill this basic necessity, and did nothing to facilitate a more widespread distri-bution of the language.

In the first phase of English in India, then, English language teaching was – for the most part – left in the hands of the missionaries. Con-fronted by what they perceived as the blasphemies of foreign gods and idols, the missionaries set out to create converts and save souls. Know-ledge of English was always seen as an essential aspect of this ultimate goal. Eventually, however, as the East India Company shifted from its purely commercial role to taking on the powers and responsibilities of colonial rule, the British saw the necessity of establishing more formal English training. It was at this point that the goals of the missionaries fused with the wider political interests of the Empire.

The explicit objective of the British was plainly stated by Thomas Babington Macaulay, an imperial official who came to India in the 1830s as a member of the supreme council of the East India Company. Macaulay argued that English teaching throughout the subcontinent would help 'create a class of English-knowing Indians who could function as interpreting buffers between the rulers and the ruled'.[26] In a famous statement, he outlined his vision of these interpreters as the perfect colonial subjects. His desire was to create 'a class of persons, Indians in blood and colour, English in taste, in opinions, in morals and in intellect'. By 1835 Macaulay's 'Minute on Indian Education' – alongside the governor general William Bentnick's 'Special Ordinance' – led to the establishment of English education in Indian schools and universities. Two years later, in 1837, English was made the official language of the government.[27]

During British rule, English in India was never widespread. At the height of the Raj, English speakers numbered only around 2 percent of the population. Yet, through a systemic and structured elitism, 'slowly', writes Kachru, 'English became the medium for the socially and administratively dominant roles . . . the legal system, the national media and important professions.'[28] In this way, the British Raj managed to create an exclusive English-speaking population that could function as Macaulay's class of 'buffers', interpreting between the rulers and the ruled.

Macaulay's dream of a population 'Indian in blood and colour but English in taste' makes it clear that the British believed teaching English would transplant not only a language but also a culture. Throughout colonialism, writes Kachru, 'English was seen as a tool of "civilization" and "light".[29] Provision of that tool is perceived as the colonizers contribution – and duty to the wellbeing of the inhabitants of newly acquired colonies.'[30] The English language was considered to carry with it Western values in religion, commerce and politics, and it was thought that, through English, these values would be brought to the subcontinent.[31]

The colonizers hoped that English would provide the linguistic and cultural means to transform Indians into loyal subjects of the British. Yet, even the British themselves recognized that there was a danger that this plan could backfire. If the English language did indeed have a culture, one had only to look at the history of Britain or, after 1783, at the even starker example of the American Revolution to realize that it was not one of passive acceptance to dictatorial control. There was always the 'risk that teaching English would lead to revolt'.[32]

By the late 1800s this risk was proving to be a reality. Lord Curzon, viceroy of India from 1898 to 1905, feared that the experiment of English education had failed 'because, if it had succeeded at all, it had succeeded in the wrong direction'. There is, wrote Curzon, a powerful school of opinion which does not hide its conviction that the experiment was a mistake and that its result has been a disaster.

When Erasmus was reproached for having laid the egg from which came forth Reformation, 'Yes,' he replied, 'but I laid a hens egg, and Luther had hatched out a fighting cock'. This I believe is pretty much the view of a good many of the critics of English education in India. They think that it has given birth to a tone of mind and to a type of character that is ill regulated, averse from discipline, discontented, and in some cases disloyal.[33]

The fear that English would contribute to a 'disloyalty' among the population of British India proved, of course, to be well founded. As Indians became comfortable with the language, they began to twist it against the colonizers, using the Englishman's tongue to bring attention to the injustices of British rule. 'There is always a difference between what message is intended and how it is received and used, and the gap quite often helps subversive forces.'[34] Vivekananda, Tagore and Gandhi all used their mastery of the English language to broadcast the 'lurid face of British imperialism' and 'beam it across the world'.[35] From Nehru's famous 'tryst with destiny' to the Indian constitution itself, in India, the language of foreign rule and colonialism was absorbed and mutated into the language of national freedom and independence.[36]

1947: Independence

Despite the fact that the 'Quit India' movement had turned the English language back on its native users, there was widespread belief that, with the end of the Raj, English in India would be replaced by one of the country's indigenous tongues. English retained a strong and sour legacy. Denounced as a symbol of oppression and exploitation, its use was invariably perceived as anti-national and anti-Indian. It was in this spirit that Nehru confidently declared that 'within one generation English would no longer be used in India'.[37]

Gandhi too was insistent that 'the question of language' be welded to 'the question of nationalism',[38] and that the strength of the Indian nation depended on getting rid of English and establishing a native national tongue.[39] Firm in his belief that 'real education is impossible through a foreign medium',[40] Gandhi regarded it as 'a sin against the motherland to inflict upon her children a tongue other than their mother's for their development'.[41] 'To give millions knowledge of English is to enslave them,'[42] he wrote, arguing that English 'should be banished as a cultural usurper [just] as we successfully banished the political rule of the English usurper'.[43]

Despite these adamant views, however, the founders of the state recognized that, out of practical necessity, English should continue to be used, particularly in the business of government, for at least long enough for the smooth transition to a native language to take place. With this in mind the constitution of 1947 recognized English as an associate official language and in Article 343 stated that: 'The English language shall continue to be used for a period of 15 years from the commencement of this constitution for all official purposes of the Union for which it was being used immediately before such commencement.'[44]

This 15-year deadline, however, has been repeatedly delayed. 'Legislation passed in 1963 postponed transition from English to Hindi indefinitely.'[45] In 1967 this postponement was entrenched in the constitution through an amendment to the Official Language Act. This amendment adopted what is now known as the 'three-language policy' in which English retains its place alongside Hindi, and both languages share their official status with a regional tongue. With the three-language policy, the position of English in India was constitutionally entrenched.

In the early years of independence English was able to retain its privileged status, not because of any positive love for the language, but rather because of the passionate opposition to anything that might replace it.

India is staggeringly pluralistic linguistically. 'The 1961 census enumerated a total of 1,652 claimed mother tongues belonging to four language families: Indo European, Dravidian, Austro-Asiatic and Sino-Tibetan, and the Eighth schedule of the Indian constitution lists eighteen languages of India.'[46] A region of such cultural and linguistic diversity needs a link language to hold the country together. This intranational language serves, among other things, to conduct the affairs of the state, to ensure equity and fair play in the all-India competitive examinations, to allow universities and research centers to share information, and to ease the flows of trade.

The reason that English is so well suited to this role is that it has been deracinated, detached from its Western roots. Having lost its overt associations with Western domination, the language is now perceived as culturally neutral. Relatively free of local biases and 'native codes of caste, religion, region' etc., English – unlike any of the indigenous tongues – was not associated with any particular region or tribe, religion or ethnicity. 'With the use of English the people are equal because they shared its advantages and its disadvantages.'[47]

This was decidedly not the case with Hindi – India's other official language, and the only real candidate for its replacement. In fact, the strength of English in independent India owes much to the strong, and sometimes violent, opposition to Hindi that arose in the non-Hindi speaking South.

In the South where each state has its own local language – many of which have long and rich traditions – Hindi is perceived as just another regional tongue. Its pretensions to be the country's official language met with strong resistance. Southerners worried that the imposition of Hindi would serve only to further consolidate the dominating rule of the more Aryan North over the more Dravidian South. They were concerned that it would give unfair advantage to Hindi speakers and unjustly favor the Hindi elite, particularly in areas of education and bureaucratic position where the all-India exams are so important. Driven by pride and loyalty to their own mother tongues, by their adamant resistance to what was increasingly viewed as 'Hindi imperialism', and by the fact that their English-speaking skills were – if the system remained unchanged – a sure guarantee to a good profession, politicians, students and others banded together to oppose the establishment of Hindi and lobby for the retention of English.

In 1956 the Academy of Tamil Culture convened in Chennai for the 'Union Language Convention'. Speakers warned against the dangers of the 'new Hindi', the imposition of which would lead to the 'inevitable disintegration of the country and the ultimate destruction of minority languages'.[48] Voicing their clear objections to Hindi,[49] the convention drafted a resolution, which stated that it would be 'greatly unjust to make any other language take the place of English, when to a population of about a hundred million ... it would be a language with which, for all practical purposes, they are totally unacquainted'.[50] This position was further reinforced at the All-India Language Conference in March 1958 where it was declared that 'Hindi is as much foreign to the non-Hindi speaking people as English to the protagonists of Hindi.'[51]

Nehru responded to the agitation that was growing in the South, and also to his own frustration at the inflexible attitudes of the proponents of Hindi, by addressing the language question in a speech to the Lok Sabha on 5 September 1959. His goal was to reassure those in the Southern states that Hindi would not be imposed against their will. 'English', he is recorded as saying, 'will continue as an associate language to the official language of Hindi, and the question as to how long English should continue as an associate language will be determined only by the non-Hindi speaking people ... I do not wish to impose Hindi compulsorily on any state which does not want it.'[52]

In 1963 the government appeared to back up this reassertion with an amendment to the Official Language Act. The amendment stated that

'English may continue to be used in addition to Hindi.' The wording of this amendment, however, was ambiguous. In particular there was a sense that the word 'may' was unclear and could be open to a variety of different interpretations. Thus, it was widely believed that when Nehru died in 1964, his personal assurances about the lack of forcefully imposing Hindi had not been made official. In the immediate post-Nehru political climate the 'idea gained ground that from 26 January 1965 Hindi was going to be the sole official language of India'.[53]

At this point the Southern opposition to Hindi turned violent. Language riots erupted in Tamil Nadu. At protests in Chennai, demonstrators carried placards reading 'Hindi never, English ever'. The protests grew more and more frenzied. Local politicians burnt themselves alive. In the end, 66 people died in agitations that lasted two months. Finally the government capitulated. English would 'continue to be used, in addition to Hindi, for all the official purposes of the Union ... and for the transaction of business in Parliament'.[54] The compromise of the 'three-language policy' was born.

The Permit Raj

Both Nehru and Gandhi had sought to assert national independence by promoting the use of an indigenous tongue. The exact same impulse governed their economic policies. Seeking to protect the country from the imperialism of the West, they promoted the independence, autonomy and self-reliance of India's indigenous economy.

In both the linguistic and the economic spheres, however, factors on the ground eventually shattered this idealism and the protectionist mindset that it entails. A sense of pragmatism had ensured that India open itself to the outside forces that it at first sought so vehemently to resist.

Nehru's initial dream was to create a 'mixed economy', superior to both capitalism and communism. Yet, his ideological commitment to the state control of industrialization ended, as Gurcharan Das writes 'in combining the worst features of both worlds – the "controls" of socialism with the "monopolies and lobbies" of capitalism'.[55] The mixed economy was not, as Nehru had hoped, a force of liberation. All it managed to do was to replace the British Raj with what came to be known as the 'Permit Raj'.

India's state-owned companies are notoriously inefficient, plagued by bloated bureaucracy and rampant corruption. The country is famous for its red tape. In the Permit Raj, these scourges on the economy were institutionalized through 'a complex, irrational, almost incomprehensible system of controls and licenses that held sway over every step in production, investment and foreign trade'.[56] This baroque and overbearing system created a stifling environment in which, for example, 'any company worth over US$20 million had to submit all major decisions, including the membership of its board of directors, for government assessment'.[57] The following passage by Das describes this Kafkaesque licensing system in nightmarish detail:

> An untrained army of underpaid, third rate engineers at the Directorate General of Technical Development, operating on the basis of inadequate and ill-organized information and without clear-cut criteria, vetted thousands of applications on an ad-hoc basis. The low-level functionaries took months in the futile microreview of an application and finally sent it for approval to the administrative ministry. The ministry again lost months reviewing the same data before it sent the same application to an interministerial licensing committee of senior bureaucrats, who were equally ignorant of entrepreneurial realities, and who also operated upon ad-hoc criteria in the absence of well-ordered priorities. Once it cleared the licensing committee, it was sent to the minister for final approval. After the minister's approval, the investor had to seek approval for the import of machinery from the capital goods licensing committee. If a foreign collaboration was involved, an interministerial foreign agreements committee also had to give its consent. If finance was needed from a state financial institution, the same scrutiny had to be repeated afresh. The result was enormous delays, sometimes lasting years, with staggering opportunities for corruption.[58]

The 'Permit Raj' was supported by mutually reinforcing groups of bureaucrats, large industrialists and politicians,[59] all of whom benefited from the insularity of the system. With time, it spread throughout the country like a cancer. Companies, protected from all competition, had no reason to innovate and Indian customers were forced to live with shoddy goods and slow, frustrating services. Combined with the deep hostility to international trade, this ensured – despite the talents of its people – that technological growth would be slow to take hold in India.

Most damning of all, however, was that these socialist 'pro-development' policies did little or nothing for those whom they were supposedly designed to help. Decades after independence, India – like other countries that had embraced the mindset and economic policies of the third world – remained mired in a terrible, debilitating poverty. As late as the early 1990s over 30 percent of the population lived below the poverty line. Only 68 percent of children were enrolled in primary school and the country had some of the worst illiteracy and infant mortality statistics in the world.[60]

1991: Reforms

In 1991 Nehru's grandson, Rajiv Gandhi, the then reluctant prime minister of India, was killed by a suicide bomber. When Rajiv's wife Sonia declared herself out of the race,[61] the Congress party was forced to look beyond the Nehru Dynasty in its search for a new leader. The party chose Narasimha Rao, not because of the qualities he possessed, but because of those he lacked. Put simply, Rao was picked because 'he was seventy, dull, and threatened no one'.[62]

Few expected Rao's government to last and fewer still thought it would accomplish anything of importance. Yet, it was the government of Rao that is responsible for what is arguably the most crucial event in modern Indian history. As Gurcharan Das writes, 'the economic revolution that Narasimha Rao launched in the middle of 1991 may well be more important than the political revolution that Jawaharlal Nehru initiated in 1947'.[63]

India's economic reforms were sparked by crises. When Saddam Hussein invaded Kuwait, the subsequent rise in oil prices shifted the Indian economy's slow motion collapse into a sudden and abrupt catastrophe. By the time Rao's government took power, the state was basically bankrupt. Foreign exchange reserves were down to two weeks of imports. 'Part of the nation's gold reserves had been flown to London to provide collateral against the 2.2 billion dollar emergency loan from the IMF',[64] NRI's were pulling out all their money from Indian banks. There were even desperate discussions about selling off the Indian embassies in Tokyo and Beijing to raise quick money.

Profound and dramatic changes were required. Luckily, Rao, through what was most likely a mixture of good fortune and foresight, had appointed a finance minister, Manmohan Singh and a commerce minister, P. Chidambaram who were up to the task. 'India needs to think afresh on many fronts,' Singh is quoted as saying soon after his appointment. 'The old methods of thinking have not taken us anywhere.'[65] As later comments by Chidambaram make clear he, too, was ready to implement a complete break with the past. 'I saw how intrusive, oppressive, and inefficient government had become, stifling entrepreneurial spirit, killing every idea, and not delivering anything in turn.'[66]

One night in late June, only a few days after the government had been sworn in, Singh, Chidambaram and a number of other technocrats began a series of meetings that were meant to address the economic disaster they had inherited. The aim was to come up with a system that would succeed in pushing the country in a completely different direction. Everyone knew that the only solution on the table was to open the country to the outside and embark, finally, on a process of liberalization and reform.

In India – as elsewhere – this process of opening up did not occur through slow incremental change but rather as a sudden and explosive transformation. The Indian economy was reformed not in months, weeks or even days. Instead, it took Rao's government just one night to kill import licensing and dismantle the country's notorious Permit Raj. In tense, closed door meetings, rules and regulations that had taken years to build up were ripped apart. In just hours 'miles of red tape, months of delays, and the hassles, anguish, and corruption that the Indian state had build up over decades' were eliminated.[67]

Rao immediately began to prepare the population for the changes that were to come. 'His government', he declared on a nationwide broadcast held only a day after taking power 'was committed to removing the cobwebs that came in the way of rapid industrialization.' Das reports that in the period immediately following, 'the government announced a new reform almost every week'.[68] The news was met both domestically and internationally with great excitement. *The Economist* magazine ran a headline story calling India an 'Uncaged Tiger' and, for a while at least, India lived up to the hype.

After more than four decades the country finally unshackled itself from what had come to be known as the 'Hindu Rate of Growth' (approximately 3.5 percent) and became one of the fastest growing economies in the world. It remains, along with China, one of only two countries that have managed to maintain a constant growth rate of above 5 percent. Reforms have also given a massive boost to India's increasingly dynamic middle class, now estimated at approximately 20 percent of the population.

Most importantly, however, the events of 1991 gave the country an overwhelming sense of confidence, which according to *Businessworld* magazine, 'transcends the world of business and pervades every area of activity'.[69] This has produced a 'tectonic shift' in the country's attitude, enabling it to shed its 'third world' frame of mind and begin openly to participate – and even compete – in the flows of global trade. Nowhere is this more evident then in the world of high tech.

English, India and IT

For over 200 years, English in India has been the language of modern technology. As authors Burde and Krishnaswamy write, 'English in India signifies technique and technology and technicality. It is not primarily a human language here.'[70]

This association between English, modernity and technological progress began during the Raj. Its force and influence was clearly entrenched by the 1800s when – in a period that scholars of English in India see as the second phase of the language's infusion into the country – locals themselves began to ask for English instruction in order that they might gain access to the modern sciences.

Raja Rammohan Roy (1772–1833) was among the most prominent of these proponents of English. Concerned that his country was being left behind due to what he perceived as 'ignorance and intellectual stagnation', Roy believed that 'the need of the hour was a rejection of medieval superstition and a grand leap forward with the modern language, literature and science of the West'.[71] He called on Europeans to teach Indians 'mathematics, natural philosophy, chemistry, anatomy and other useful sciences, which the natives of Europe have carried to such perfection'. His views were dramatically condensed in a letter of protest to Lord Amherst dated 11 December 1823, on the occasion of the establishment of the Calcutta School of Sanskrit. Voicing strong opposition to the idea that state funded education should focus on 'oriental learning', Roy called for a school that would concentrate its teaching on the science and literature of Europe. 'It would be foolish', wrote Roy, 'to load the minds of youth with grammatical niceties and metaphysical distinctions of little or no practical use for possessors or to society.'[72] Roy's efforts eventually resulted in the establishment of the Presidency College of Calcutta.

Later in the 20th century, Nehru also strongly associated English with both techno-science and modernity and forcefully argued that the language had to be maintained in order to preserve India's link with the outside world. He warned that to discard English after the British left would 'amount to closing a window on the world of technology'.[73] 'English is important because it is the major window to the modern world for us,' he is quoted as saying. 'That is why we dare not close that window. If we close it, it is the very peril of our future.'[74]

Today, this link between English and the flows of trade and technology is even more strongly reinforced because in India – as elsewhere – English is the language of IT.

Though it is extremely difficult to gain accurate statistical information, most analysts guess that the number of English speakers in India hovers at around 5 percent of the population. Though this may seem like a tiny fraction, in a country the size of India it is still a vast amount. Five percent of the population is equal to approximately 50 million people. This means, according to some 'guesstimates', that 'more Indians speak English and write English than in England itself'.[75] In addition, India's English-speaking population are among the best educated in the country. Many of them pursue science and engineering degrees where 'English continues to be the only medium of teaching and examination'.[76] These combined factors have allowed India to become home to the second largest pool of English language engineers in the world.

India's English-speaking skills give it tremendous advantages in the world of IT. First, because English has, at least historically, been the dominant language of digital technology. ASCII, traditionally the most common encoding scheme, used on the majority of the world's PCs and on both Unix and DOS operating systems,[77] has the capacity to represent only 256 characters.[78] Though this covered some – but not all – of the European scripts, it meant that, at this fundamental level, computers could not read India's indigenous tongues. Furthermore, though computers are instructed in machine code, which is comprised of nothing but zeroes and ones, programming languages, the interface between humans and machines, are predominantly rooted in English. Some argue that these languages are so different from normal English, using nonsensical terms and assigning different meanings to well-known words, that English language users have no particular advantage when it comes to programming. However, many programming languages use at least some familiar terms such as PRINT, READ and STOP and almost all are written using the English script. This combined with the fact that most manuals, web pages and Internet communities designed to help programmers are in English, ensure that even today English retains its primary position.

More generally, India's linguistic advantage has helped its scientists and engineers to be directly plugged in to the – overwhelmingly English – world of science and technology. The majority of scientific books and journals are in English and 'over two-thirds of the world's scientists read in English'.[79] Due to their language abilities, Indian students and researchers are able to read and contribute to the latest developments in their field without having to wait for translations. Knowledge of English has also enabled those who travel or move overseas to have a relatively easy time assimilating into the scientific and technological communities they find abroad.

This ease of integration has been crucial for the IT industry as a whole. The large pools of English language engineers allowed local companies – as soon as the protectionist barriers were brought down – to seamlessly merge with global businesses in a way that only a common language makes possible. The medium of IT, the Internet, is still predominantly English. Of the estimated 200 million users of the Internet, over 35 percent communicate in English.[80] The figures for web content are even more dramatic: Of a total of 313 billion websites, 68.4 percent are in English. The next largest linguistic group is Japanese, which comprises only 5.9 percent.[81]

In the software industry, trends from body shopping, to outsourcing, to the establishment of offshore development and call centers have long capitalized on India's enormous pool of English speakers skilled in technology.

There are those like Trivedi who argue that the fact that India's growth in IT is so bound up with the English language is evidence that the country's involvement with high technology is intimately bound to the long, ongoing process of Westernization. Yet, English, like IT, is not inextricably tied to the West. Once opened up, the networks of globalization are influenced as much – if not more – by the creative mutations that seep in from the periphery.

3
Eastern Influences

They called it the Asian Economic Miracle because the world
had not really seen that kind of economic growth, that many
people brought out of poverty, that rapid a creation of the middle
class anywhere in the history of the world.
  – Daniel Yergin, *Commanding Heights: The New Rules of the Game*

Summer 1999: Jubilee Hills on the outskirts of Hyderabad

S. Hariharan, deputy manager from L&T Infocity, sits in a fully wired
office, behind an imposing black and shiny desk. The office is on the first
floor of 'Cybertowers', a vast circular edifice made of brick and mirrored
pane. From the tea shack across the road, where men perch themselves
on rickety stools sipping *chai*, the building appears like a desert mirage.
Rising anomalously from empty surroundings, it seems to belong more
to the science-fiction landscape of Hong Kong or Singapore than to the
barren hills of Andhra Pradesh.

Mr Hariharan, however, is indifferent to the apparent incongruity.
In a nonchalant tone that exudes corporate confidence, he gives a brief
introduction to his company's role in constructing the future of India.
When he is done, he summons a subordinate who will act as our guide.
The building is barely complete, but already its floors are occupied. As the
elevator climbs, it passes the offices of some of the high-tech world's most
famous MNCs – Microsoft, Oracle, GE.

The tour culminates on the building's rooftop. From here it becomes
clear that 'Cybertowers' is only the first step in a much wider and more
ambitious plan. The guide gazes out onto the shrubs and strange rock
formations of the Deccan plateau. Pointing into the distance he states,

with a voice of absolute certainty, 'the hotel will be there, the shopping complex here, the hospital over there...'

Construction has already begun, but it is occurring in a most surreal manner. Though the aim is to create 'a self-reliant Infocity' with 'state-of-the-art' infrastructure and 'every conceivable amenity', 'Hi-Tec city'[1] is being built in a typically low-tech Indian manner. Cranes and bulldozers are nowhere to be seen. Instead, workers balance precariously on bamboo scaffolding, while, on the ground below, men with simple tools crush stone by hand and women in saris move vast piles of rocks on their heads.

The almost Pharaonic scale and grandeur of the project is undoubtedly impressive. Yet, one is also invariably struck by a sense of temporal jarring. How can something so futuristic be emerging from so archaic a place? India, like most 'developing' societies, has a kind of anachronistic culture in which the deep past mingles easily with the far future. Time here is often scrambled, and change rarely occurs through gradual steps or stages. Rather, one moves – in an instant – from the tea shack on one side of the road to 'Cybertowers' on the other. Here, on the periphery, the future does not emerge through a slow and steady process of development but arrives instantaneously through discontinuous jumps.

The gunas revolve

In his book *Yoga: Immortality and Freedom*, Mercea Eliade explains the Indian doctrine of the gunas by writing that 'prakriti' or 'primordial substance' has three 'modes of being', which enable it to 'manifest itself in three different ways'. These 'modes of being' or 'gunas' are 'sattva (modality of luminosity and intelligence); rajas (modality of motor energy and mental activity); and tamas (modality of static inertia and psychic obscurity)'.[2] All three gunas exist simultaneously in any given time and place. Yet they do so in unequal proportion such that one guna always dominates. This domination, however, does not last forever. One of the fundamental principles of Hindu philosophy is that the gunas revolve.

Ten years after the 1991 reforms there was a gathering sense in India that the accomplishments, dynamism and period of growth that the country had recently experienced were all the result of a single shift that had occurred over a decade ago. Accompanying this realization was an ambient anxiety, a feeling that momentum had been lost and that a period of stagnation was setting in. 'Everyone is concerned with the question' said Sam Pitroda, addressing an audience at TiEcon New Delhi 2003, 'Why can't we grow faster?'[3]

This atmosphere of concern was rooted in the fact that the 7–8 percent growth rate that the country had enjoyed in the early to mid-1990s had more recently slipped down to hover around 5 percent. Moreover, though everyone agreed that another boost was needed and that 'second-generation reforms' – as they had come to be known – were long overdue, no government showed any signs of having the courage to initiate this process.

The overall impression, then, was that the reform process had stalled, and that, as India entered the 21st century, 'Nehru's tryst with destiny remained a promise unfulfilled'.[4] Tamas, which Sri Aurobindo called 'the dark and heavy demon of inertia', was threatening once again to take hold.

Yet, traveling in India in 2003, one also sensed an increasingly strong and articulate counter-tendency that sought to oppose this trend. In a *Businessworld* round table, conducted in association with TiE, moderator Tony Joseph began:

India has had two periods of heightened expectations. One was in the early 1950s, in the first flush of independence. Nobody at that time would have doubted that 50 years later, India would be a developed country. But we took a wrong turn and ended in a blind alley. The second period of heightened expectation was in the early 90's, when we broke from the shackles of socialism. But that dream also died soon. Today, I would say we are again approaching a period of heightened expectations. I would assert that there is a sense of national belief that India is destined for higher levels of growth.[5]

At the forefront of this new wave of expectation lay a group of, mostly American, business leaders. People such as Rajat Gupta, managing director of McKinsey; C.K. Prahalad, professor and business consultant; and Kanwal Rekhi entrepreneur and angel investor, all of whom had become intensely involved in the contemporary development of India.

Traveling frequently throughout the country, attending conferences, addressing the media and consulting with industry and government, these men repeated the same message over and over again until it took on the strength of a mantra. Rapid advance is an imperative for India, they claimed. For the country really to prosper it is mandatory to achieve an annual growth rate of 10 percent.

For this goal to be achieved, the country will have to adopt many radical and far-reaching transformations. These range from technocratic policies (streamlining taxes, removing import duties, reforming labor laws) to more widespread cultural shifts like curtailing corruption. Yet, despite the scale of the changes required, there is a widespread belief that they will not occur gradually. India's second-generation reforms must arrive – like its first – as a sudden and revolutionary break.

'Though Indians may be poor,' says C.K. Prahalad in his speech at TiEcon New Delhi, 'India is a rich country, not a poor country.' He illustrates his point by focusing on a single issue. A common complaint is that India suffers from a resource constraint, he says. Yet, India has more resources than it knows what to do with. The country's rice stock, for example, is more than the total world trade of rice. Its wheat stock is more than three times the world trade of wheat. The problem is not a lack of resources but a misallocation of resources, trapped resources and wasted resources. 'Everyone in the country', Prahalad insists, 'knows the problem and also the solution.'[6]

As Prahalad speaks, one gets the impression that he is evoking an altogether different country, a kind of 'virtual' India – wealthy and prosperous – that exists on the other side of an invisible barrier. Though it seems, at times, impossibly far away one senses that this virtual India can only be made manifest if it is grasped all at once. If this were to occur, if the gunas again were to revolve, India could experience another profound and dramatic discontinuous break, and, in the space of only a few years, unleash itself once again.[7]

Looking East

The model for rapid development does not, of course, come from the West, where 10 percent growth is unheard of. Rather, the confidence that – given the will – India's growth rate could explode exponentially comes from the impact of the East Asian example.

There is no doubt that India's Eastern neighbors influenced its first wave of reforms. In 1987, Manmohan Singh visited the region and was stunned by what he saw. The comparisons with India were astounding. Half a century before, the countries of East Asia were as poor as Africa is now. In Taiwan, parents sold their children to keep from starving. At the beginning of the 1960s South Korea's GDP was lower than India's while Japan's was not much higher after World War II. By the late 1990s, however, the region had completely rebuilt itself. The Japanese were among the richest population in the world and per capita income in Taiwan and Korea was at least 'ten times that of India's'.[8]

The primary reason that East Asia did not languish in a state of underdevelopment – despite the fact that it is often regarded as being located in the periphery – is because of the attitude it adopted to the outside world. Seeking intimate connections, rather than protective enclosures, the countries of East Asia aimed to insert themselves directly into the flows of global trade.

Their strategy was to focus on low cost manufacturing and, particularly after the 1960s, on the mass production of electronic goods. Unlike India, they did not aim for self-sufficiency but rather welcomed, even encouraged, foreign investment. Taiwan's television industry, for example, was made possible by investment from companies like Phillips and Mitsubishi. In this way, the Asian Tigers created what has come to be known as the 'export-oriented development model', a paradigm of growth that is based on the production and export of low cost goods.

This model fundamentally challenged the modernist notion that technological progress proceeds through a series of developmental stages which imitate – or at least conform to – a Western archetype. The countries of East Asia operated with the realization that, as the world globalizes, interlocking parallel developments have replaced relations of mimicry and succession.[9] Ceasing to view their development as bound to the slow gradual stages of historical time, they were able to take advantage of this new global simultaneity. By adopting a strategy which relies less on state intervention and more on global trade, Asia now seeks to compete – rather than to play catch up – with the West.

Their success has been staggering. The Asian Tigers have not only managed to lift themselves out of destitute poverty, they are now poised at the future's cutting edge. From the latest in robots to wireless, commodities that are still considered to be in the realm of fantasy in the West, barely get a glance in the East. It is common practice now for high-tech companies to test their new products on the streets of Tokyo or Hong Kong before releasing them in America.

Asia has the largest productive population on the planet. Its cities combine skyscrapers and futuristic gadgetry with the frenzied activity of 'wet markets'.[10] These factors combine to ensure that the near future belongs to Asia. Speaking at the Indian IT Forum in Hyderabad, Tharman Shanmugaratnam, a minister from Singapore, made this point by stating the following: 'In the past century no Asian city has emerged as a leader in the digital economy. Asians were leaders, but they lived in American cities. The next phase will belong to the Asian cities – Bangalore, Hyderabad, Singapore, Shanghai...'[11]

Shanmugaratnam is a diplomat, and he was speaking in Hyderabad, but the truth of the matter is that the extent to which India will actively participate in the ongoing rise of Asia is still uncertain.[12] What is beyond doubt, however, is that if it is to achieve its aspirations and play a decisive – even leading role – in the coming 'Asian century', India needs to shift the direction of its gaze.

For centuries the country has been geared toward the West. Partly this is because the two latest waves of ruling invaders – the Mughals and the British – come to India from the West. Yet, India's Western fascination did not end with independence. On the contrary, independent India's elite and intelligentsia were completely focused on Europe and later America. That is where they went to do business, to travel and to get educated. The result was a complete obliviousness to what was going on next door. This is why, according to Professor Anand Patwardhan of IIT, for India the rise of the Asian Tigers came as an almost total surprise. 'Suddenly we realized something phenomenal is going on.'[13] Professor Patwardhan is optimistic, however, that India is now beginning to realize that it 'needs to have a much more eastward looking frame of my mind – not just in economic terms but also in cultural terms. My hope', he says, 'is that in the next decade we will go through this process of rediscovery – of Thailand, of Cambodia, of Vietnam – of the whole of South East Asia.'[14]

The dragon next door

India's more eastward looking focus has forced it to come face to face with the rapid rise of its giant neighbor. Though the emergence of the Asian Tigers challenged protectionist models of development, the example they provided was easily dismissed inside India as irrelevant. 'Indians coming into the 90's', explains Kanwal Rekhi, 'thought, you know, Korea, Taiwan, Malaysia, Hong Kong these are small countries. But how do you develop this giant. It wasn't until China started to show 10 per cent growth that we started to say hmmm.'[15] For India the only real model of development is China. 'When you are a country of a billion people', argues Madanmohan Rao, 'the only other benchmark can be another country of a billion people.'[16]

The mere fact of China's growth, then, acts as a wake-up call for India. 'China holds a strange fascination', claims Rao, 'because to see another country like us grow so fast and so spectacularly has just left everyone spellbound.'[17] This is why someone like Kanwal Rekhi, who works hard to challenge the sluggishness of India's status quo, insists on mentioning China in all his speeches.[18] As C.K. Prahalad says, the message China gives is obvious. 'Now we cannot say we can't do it (grow at 10 per cent) because someone – China – has been demonstrating for 20 years that it can be done.'[19] 'China's ability to grow rapidly and reduce poverty', concurs economist T.N. Srinivasan, 'forces those inside India to ask why it is that they can do it and we can't.'[20]

By 2003, one had only to step foot in India to realize that this message had sunk in and that China was on everybody's tongue. Newspaper articles, magazine columnists, government officials and private entrepreneurs were all obsessed by the dragon next door. A quick glance at statistics shows why. In the late 1940s, at the beginning of the modern period for both China and India, the two countries were at a comparable level of development. 'Both countries', writes Amartya Sen, 'were among the poorest in the world and had high levels of mortality, undernutrition and illiteracy.'[21] Sixty years later the 'Chinese are a generation ahead'.[22] While China, in 2002, had 40 million people living below the poverty line, India still struggled with 400 million. Despite impressive growth in the telecommunications sector, India had only 10 million mobile phones compared with China's 200 million.

Though they are both equally rich in terms of cultural and historical treasures, China attracts approximately 86 million tourists annually while India has a tiny 2.6 million visitors every year. 'Infant mortality is more than twice as high in India (71 per thousand live births in comparison with China's 30).'[23] China has 5.4 percent of world trade while India has a mere 0.63 percent. India has 1.35 percent of World GDP while China has more than twice that at 3.2 percent. India's exports amount to 44 billion dollars while China's exports amount to almost ten times as much (at 421 billion dollars). Perhaps most striking of all, India attracts 3.5 billion dollars in foreign direct investment, compared to China's 105 billion dollars.[24]

One commonly cited explanation for this staggering divide is that China's reforms, which began in 1979, are 12 years ahead of India's, which only began its reform process in 1991. There is a widespread fear in India, however, that despite the process of liberalization, the gap has continued to grow, and that, rather than starting to close, it threatens to keep expanding exponentially. This is especially true if China keeps growing at 10 percent and India's growth rate lingers in the 5 percent range.[25]

China, then, acts as a crucial motivation for India's development, shattering excuses and providing a positive model to follow. The lessons it offers are many and varied. One of the most fundamental has to do with the importance of basic infrastructure. In China, electricity, transportation and communication all work smoothly and efficiently. India, on the other hand, is notorious for its potholes and power outages. In New Delhi, for example, it is not at all uncommon to be plunged into darkness as frequently as twice a day. This difference alone has huge effects on productivity and foreign investment (not to mention quality of life) and – besides general levels of economic freedom – is probably the single most important factor in the gap between the two countries. Another crucial difference is that China's growth rests largely on manufacturing, a sector which comprises 35 percent of China's GDP. In India, on the other hand, manufacturing remains strangled by bureaucracy and red tape and comprises only 15 percent of GDP.

There are signs that some of these lessons are getting through. The Indian government, especially in certain states, is beginning to pay much more attention to basic infrastructure, recognizing its crucial importance to all other elements of growth. At the same time, many within the country are advocating a renewed stress on manufacturing, lobbying hard for the required reforms that can make this possible. The country has even begun experimenting with Special Economic Zones that emulate areas like Shenzhen in China by focusing on the manufacturing sector. These zones, writes the *Far East Economic Review* offer 'generous tax breaks, simplified customs and foreign investment procedures, improved infrastructure and an unusually helpful bureaucracy'.[26]

Yet, despite the importance of these specific issues, what is far more critical is the simple fact of China's awesome growth, and the perception, within India, that it happened all at once. *Businessworld* magazine, for example, reported that in China, '95 per cent of national wireless telecom infrastructure was put up within three weeks'.[27] Whether this is true or not, it feeds the impression in India that it is possible for sub-stantial – even fundamental – change to occur in sudden leaps. It is this belief, which awes and inspires the country, helping to create the over-whelming impression that the next decade is a 'window of opportunity' for India. If anyone doubts this, says Sam Pitroda, one of the country's great reformers, 'they have only to look at what China has done in the last two decades'.[28]

China's growth, which in 2002 amounted to 15 percent of world economic growth, and nearly 60 percent of the world's export growth, does evoke, in some, feelings of anxiety and resentment. Suspicions of world trade and fear of competition converge to create the worry that floods of cheap goods from China will kill local industries.

Yet, on the other hand, there is an increasingly widespread awareness, at least in Asia, that despite these concerns, the globalized economy – especially in the information age – is characterized less by inherent structural inequalities and more by interdependence and non-zero sum games. For those who hold this view, China's growth is something to celebrate. 'It is time', writes Niranjan Rajadhyaksha, expressing this opinion, 'to stop talking about the Chinese threat and start talking about the Chinese opportunity.'[29]

India, as Rajadhyaksha points out, 'is well positioned to take advantage of the changes in China's economy'. China is 'focusing on its domestic market to fuel future growth'.[30] It has an expanding middle class who now consume US$244 billion worth of imports. Even more importantly for India, China's growth is increasingly being fuelled by electronic manufacturing and exports. At the same time, its own population are avid consumers of IT. The privileged one-child generation has grown up with video games, the Internet and mobile phones. By 2003 the Chinese IT market, at approximately 30 million people, had become the second largest in the Asia-Pacific region after Japan.

All this hardware needs software, and Indian software companies have already set up development centers in places like Shanghai to try to get close to the market. Their hope is to shift the relationship between India and China from one of suspicion and rivalry into one of mutual benefit and cooperation. In the most optimistic scenarios the two Asian giants will fuel the future of the digital economy by feeding off their immense pools of complementary skilled labor. In this vision China will become the manufacturing capital of the world, while India emerges alongside it as the world's software and services capital. As Zhu Rongji, the then prime minister of China, said on his visit to India in 2002, 'you are the first in software and we are the first in hardware. When we put these together, we can become the world's number one.'[31]

The rise of China, then, provides both an example and an opportunity for India. Yet, it raises important and troubling questions – not only for India, but for the world. These arise from the prevalent assumption that it is democracy that is to blame for India's failure to compete with China, and that the speed and extent of China's development can be attributed to the efficiency of its authoritarian rule. Political freedom, it is commonly believed, slows the pace of economic growth. 'You can't have economic freedom, social freedom, and political freedom, at the same time,' says Sunil Mehta of Nasscom, expressing this widely held view. To the question: 'Is the price of political freedom lower economic growth?' Mehta replies without a shred of doubt, 'not lower economic growth, but slower economic growth'.[32]

In most of the world, economic freedom preceded political freedom. In India, however, the opposite is the case. 'India embraced democracy before capitalism,' writes Das. 'This makes its journey to modernity unique.'[33] The strength and vibrancy of Indian democracy is praised throughout the world. In their award winning book, *Commanding Heights*, Daniel Yergin and Joseph Stanislaw write:

> India's commitment to democracy stands as one of the great achievements of the second half of the twentieth century. Its free elections, independent judiciary, free press, and free speech were in marked contrast to political realities both in its region and in much of the developing world, which succumbed for long periods to dictatorship, ethnic wars, and political fission.[34]

Nevertheless, India is now seen to be paying a substantial price for the freedom it enjoys. Political freedom, with its free flows of information and its corresponding right to free expression, has, it is claimed, hidden costs. Freedom of expression – it is generally believed – inevitably hinders progress. When the Chinese government wants to build a road, it just does it. If India wants to do the same thing, however, it is met with protests that must be negotiated and resolved. Inside India this 'price' of freedom has been labeled the 'democracy tax'.

However impalpable, there may be some truth to this idea. Authoritarian control appears to come with the ability to implement change quickly and efficiently. Nevertheless, this line of thought is sufficiently flawed that it is still possible – without simply appealing to idealism – to agree with Amartya Sen when he writes: 'The fact that India's record is terrible in many fields where China has done quite well does not provide a good reason to be tempted by political authoritarianism.'[35]

First, because the idea of a democracy tax credits China's biggest weakness (its repressive and authoritarian regime) for its greatest strength (its vibrant and relatively open economy), while blaming India's greatest strength (its political openness and defense of personal freedom) for its biggest weakness (its closed and bureaucratic economy).

Secondly, the notion that it is the right to free flows of information and free expression that hinders progress sidesteps the real problem with democracy, which is not political freedom but populism. All democratic governments – even those that fervently support free trade – regularly sacrifice economic freedom in order to secure voting blocks and cater to entrenched special interests. Speaking at TiEcon, New Delhi 2003, R.R. Shah, Secretary in the Indian Department of IT, was refreshingly candid about this. When a member of the audience asked why reforms could not happen faster, Shah replied bluntly because 'good economics does not always equal good politics'.[36]

Why, in democracies, bad economics equals good politics is one of the most crucial issues of our times. The answer, no doubt, is partly because in the process of liberalization and opening up, the losers inevitably feel the effects much more severely than the winners. Power reforms in India, for example, would dramatically benefit the entire population since the country suffers few greater impediments to growth than its constant power outages. Nevertheless, any mention of such reforms draws loud and sometimes violent protests from those who are currently receiving highly subsidized – or even free – electricity. In order to signal the importance of this issue Gurcharan Das chooses for an epigraph to his book *The Elephant Paradigm*, which seeks to deal with India's contemporary struggle with change, a quote from Machievelli's *The Prince*:

> There is nothing more difficult to arrange, more doubtful of success, more dangerous to carry through than initiating change... The innovator makes enemies of all those who prosper under the old order, and only lukewarm support is forthcoming from those who would prosper under the new.[37]

This is perhaps why it is extremely rare – if not unheard of – for a democratic government really to try to sell economic liberalism. Yet, without this sales job reform is impossible. It is for this reason that Kanwal Rekhi insists that India 'has not even started reforming [since it has] not even started to say growth is good per se'.[38]

Tavleen Singh, columnist for *India Today*, has written frequently on this topic. In an article entitled *A Freedom Foiled*, devoted to sharp criticism of the idea of the democracy tax, Singh rails against the idea that the Indian leadership has pushed hard for reforms, only to be curtailed by the people:

> Democracy is about the people's will and when have the people ever objected to a road being built? Have we ever heard of protests against a school or a health center being opened in a village? Have ordinary Indians ever demanded that they continue to be forced to live like pigs instead of with modern standards of sanitation and hygiene...[39]

When considering this issue it is illuminating to compare the leadership of Narasimha Rao with Deng Xiaoping. Deng is the single person who is most responsible for China's reform policies. In order to carry them through, Deng not only had to convince the vast machinery of the communist party to abandon its ideology and prejudices, he also worked hard to convince the entire population – which had been steeped for 30 years in Maoist doctrine – to completely change its everyday practices. He did this, in part, through a mass publicity campaign. The slogans of this campaign, 'Reform is China's second revolution', and 'Poverty is not socialism. To be rich is glorious' are famous throughout the country and known throughout the world. Still today, China's communist party does more than probably any other government to sell reforms to its population, devoting, for example, countless television shows and newspaper articles to the benefits of globalization and the WTO.

Rao, on the other hand, did not really bother to explain what he was doing, and did not try to sell – or even defend – his reforms. He behaved, writes Das, 'as though he did not believe in his revolution'.[40] It was clear that 'no one in the political class – neither bureaucrats nor politicians were truly enthusiastic about the change'.[41] Despite the fact that 'the nineties had been the best years in India's economic life', and that Rao's reforms 'delivered outstanding macroeconomic growth, and a sense of confidence and excitement'[42] the government was completely incapable of capitalizing on its success.[43] 'No one in India celebrated the reforms. Rao did not stand tall and take credit for historic achievement. And the people responded with a verdict that booted his party out.'[44]

The most profound and crucial question that China's growth poses to India – and to the world – lies ultimately in this contrast. How is it that as democracies turn to protectionism and populism, which benefit the few at the expense of the many, it is left to an authoritarian 'communist' government to try to sell the benefits of a market-based economy? Why is it that the communist party of China, rather than a democratically elected government, is the only government to base their political power on the success of reforms, and on rapid and sustained economic expansion?

Chandrababu Naidu – the CEO of Andhra Pradesh

To call India a land of contrasts is a tired cliché. It is the strength of clichés, however, to harbor great insight. So it is the case that if India proves a rule it also tends to provide the exception to that rule. It is in India where the relationship between democracy and capitalism, political and economic freedom seems most troubling. Yet, it is also in India where one finds the strongest example of a democratically elected politician who bases his power firmly on a platform of liberalization and economic growth.

Chandrababu Naidu became the general secretary of the Teluga Desam Party in 1985. Ten years later he was elected the chief minister of Andhra

Pradesh. Since that time he has devoted himself to propelling his state – and especially its capital city, Hyderabad – into hypergrowth.[45]

Naidu's strategy is to create a situation in which economics drives politics. He operates with the belief that 'as long as he can continue to translate his economic vision into reality people will continue to vote for him'.[46] 'Naidu is showing how a reformist vision based on conviction and direction can actually work. His tenets of development are highly effective: they appeal to pragmatism, not populism.'[47] Andhra Pradesh was the first state to directly negotiate a World Bank loan and is leading the way in disinvestments and privatization. The stated aim of the government is to achieve a GDP growth rate of 9–10 percent.

Naidu recognized very early on that his first step would have to be to try to sell reforms. In an article on global cities in *National Geographic* magazine, in which Hyderabad was highly praised, Naidu's secretary, P.K. Mohanty explained that this sales job required, in the first instance, 'a visual process'.[48] The government thus set out to transform Hyderabad, in order, says Mohanty, 'to show that something is happening'.[49]

Naidu's team embarked on a range of tasks from fixing basic infrastructure, to a 'greening' of the city, to restructuring city services. It invested in a number of feel-good projects including the construction of entertainment parks, amphitheaters and sports stadiums. Many of these were done through privatization and all were aimed at proving that, in India, what is needed is not money but reforms. 'Large cities of the third world', says Mohanty, 'are reservoirs of wealth...Andhra Pradesh isn't richer than other states it's just better governed.'[50]

The external transformation of Hyderabad worked to create an enormous shift in people's mindset. Indian businessmen who had fled to America, and, even ten years ago, would not have even dreamed of living in Hyderabad, were returning to the city. This optimism seemed to spread to all sectors of life. Inside a newly built stadium in Jubilee Hills, a basketball coach trains his team for the upcoming national competition. When asked what he thought about his chances, he answers with confidence, 'the AP team has improved remarkably in the last 5 years', he says. And then, after a short pause, 'everything has improved thanks to the visionary leadership of our honorable Chief Minister'.[51]

Naidu's development strategy, which resonates with the rest of the country, is to stimulate the economy by focusing on certain high growth sectors. In particular, he is relying on an almost obsessional commitment to IT. Nicknamed the laptop minister, Naidu is famous for converting all of his ideas into Power Point presentations, and insists on regular video conferencing with all his staff. His government has implemented a host of IT-friendly policies that range from encouraging software development to making the state an easy place for outsourcing. Naidu has modified existing laws to suit the requirements of the industry, allowing the employees of IT companies to work around the clock and – much more controversially – changing labor laws so that women would be free to work at night.

Naidu's aim is to replicate Bangalore's success, and he has succeeded in making Hyderabad one of the most important IT centers of the country. In 1998 Naidu pulled off an early and impressive coup when he managed to convince Microsoft to set up in Andhra Pradesh. Since then Hyderabad's software export industry has grown at an average of more than 140 percent annually. Nasscom has ranked Hyderabad the top IT destination in the country. 'I do not have a shred of doubt that the future is ours', said Naidu after his return to power in 1999. 'Andhra Pradesh is ready to take off and become a superpower of information technology.'[52]

Naidu's commitment to IT extends beyond attracting foreign investment and expanding software exports. He is also at the forefront of implementing 'e-government', and has developed a host of initiatives that use IT to deliver citizen services. These include: CARD (Computer Aided Registration Department), which automates sales registration, cutting processing time from 10 days to an hour; FAST (Fully Automated Service for Transport), a computerized service for drivers licenses and vehicle registration; MPHS (Multi-Purpose Household Survey), a vast databank containing a total of 76.5 million records, which citizens can use to access crucial documents such as caste certificates, birth certificates and land records; and e-Seva, a 'one-stop shop for citizen services' where people can pay utility bills, make tax payments, and apply for certificates, licenses and permits.[53] All these services are highlighted on AP online http://www.aponline.gov.in/, an internet portal set up as a joint project between Tata Consultancy Services and the government of Andhra Pradesh.

Though e-government is spreading throughout the world, nowhere is it more important than in a place like India, where people spend an inordinate amount of time and money dealing with even the simplest government services. Paying a bill, getting a license or visa, buying a train ticket, even making a simple bank deposit often involves waiting in long lines, dealing with numerous bureaucrats and moving from office to office, all of which operate with inconvenient business hours. Offering the same services quickly, simply and cheaply through IT may, as *National Geographic* writes be 'old stuff in the first world', but it is 'revolutionary here'. This revolutionary potential is further reinforced by the fact that e-government tends to counter corruption. 'Technology is the best solution for corruption', claims Mohanty, 'since when you eliminate people based transactions, a lot of corruption is eliminated.'[54] Yet another side benefit is that e-government has the ability to give access and training to huge sectors of the population who are otherwise totally unfamiliar with IT. It is, thus, a crucial tool in bridging the digital divide.[55]

The centerpiece of Naidu's vision, however, is undoubtedly Cyberabad, a city district of which 'Hi-Tec city' is only one part. Naidu's goal is to make Cyberabad into a third pole capable of complementing the twin cities of Secunderabad and Hyderabad that constitute the state's capital.

Driving through Jubilee Hills in 1999, Cybertowers was the single modern building in an area filled with scrub bush and boulders. Returning to visit in 2003, however, it is clear that this was only ever Phase 1 of the plan. Having achieved 100 percent occupancy in 14 months, Cybertowers is now home to Microsoft, Infosys, Oracle, GE and HSBC. 'Cybergateway' – Phase II – is an 866,000 square foot complex built over 8 acres of land that was completed in July 2002. It now houses the offices of GE, Oracle, Dell, Satyam, Microsoft and the STPI and has also reached 100 percent occupancy. Construction has already begun on Phase III – 'Cyber Pearl'. Built as a joint venture between L&T Infocity and Singapore-based Ascendas, Cyber Pearl is expected to be complete in the first quarter of 2004.

These mega projects are complemented by a host of other private companies, educational establishments and entertainment complexes that are springing up in the area. A tour leads one past a recently opened Infosys campus and Satyam call center, a sports stadium that just held the national games, a campus of IIIT, one of the most prestigious IT education and training centers in the country, and the International School of Business, which is partnered with the top three international business schools: Kellogg, Wharton and the London School of Business. This institution was originally to be set up in Mumbai, but Naidu managed to poach it at the last minute. The tour ends at Hitex, a mammoth exhibition hall and conference center aimed at attracting high-profile trade fairs and industry events.

This pace of change shows no sign of slowing, with hotels, apartments and recreation areas already being built. Projects in the pipeline include a hardware park, a biotechnology park and an international airport connected to Hi-Tec city by a four-lane expressway. Inspired by Beijing's successful bid for the 2008 Olympics, Naidu was even rumored to be considering making a bid to host the Olympics in 2012.

The most famous site in Hyderabad is the Islamic monument known as the 'Charminar', which was built in 1591 and is composed of four intricately carved minarets. Located in the center of the town it is surrounded by the narrow lanes of an old bazaar that have undoubtedly changed little over the centuries. It seems to be separated from the Cybertower rooftop by centuries. Yet, today the same people that visit this souk now have access to Cyberabad. This is pure future shock, and it is what accounts for the feeling that in Hyderabad – like in other cities of the East – anything is possible.

4
Marginal Capitalisms

> Every advance of culture commences with a new period of wandering.
>
> – Carl Bucher quoted in *Tribes*

14–15 June 2002

Thousands gather at the Westin Santa Clara in Silicon Valley for what is billed as the largest entrepreneurial conference in the world. For two days speakers, panelists and delegates speculate on the latest trends in technology, debate the opportunities and challenges of innovation, and discuss practical issues raised by the globalization of the IT economy.

Keynote speakers include Narayana Murthy, founder of Infosys, one of the most valuable software companies in the world; Vinod Khosla, co-founder of Sun and partner at Kleiner Perkins, a major Venture Capitalist firm in the Valley; Myron Scholes and Martin Perl, two Nobel Prize winning professors from Stanford; White House advisor Glenn Hubbard; and spiritual leader Deepak Chopra. Panels are filled with executives from such companies as Microsoft, Sun, IBM and GE. The audience is comprised of a whole range of entrepreneurs and professionals. Parallel tracks are geared toward companies of all sizes, from the start-up to the global corporation, and special interest groups narrow in on such topics as software, networking and biotech.

A year after the bursting of the tech bubble, 10 months after 9/11 and in the wake of the Enron scandal (with the WorldCom fiasco about to explode), the conference is overshadowed by the economic downturn. A number of delegates comment on the downbeat mood, comparing the event to previous years when, at the height of the high-tech boom, multimillion dollar deals were being made around every corner.

Nevertheless, from the perspective of an outsider at least, the event has awe-inspiring energy and a contagious entrepreneurial spirit. Even in the downturn, Silicon Valley – despite the prefab suburban feel to the place – has mystical allure. This, after all, is the cutting edge of what is still the strongest economy in the history of the world. The CEOs, CTOs, Venture Capitalists, entrepreneurs and professionals who have congregated here are a hub in a network that is propelling America's techno-economic power into the realm of science fiction.

What is striking is that, though this event is taking place at the heart of Western power, in the center of the global economy, most of the people gathered here would be just as at home in Asia as in America. Though, they can only be considered 'insiders', there is barely a white face in the crowd.

The conference is being hosted by an organization called TiE, an acronym for The Indus Entrepreneur.[1] TiE was founded in 1992 when a small group of successful entrepreneurs, senior professionals and corporate executives met at the airport while waiting for the delayed flight of a dignitary from India. It soon became apparent that, though they were each thriving individually, they had no collective strength. Unlike other ethnic groups, Indians in the Valley did not yet form a web. The group agreed to create a space for informal gatherings where they could share their experience and contacts. The explicit initial aim was to use their common culture as a starting point for networking so the community as a whole could benefit.

In the decade since, TiE has become immensely successful, swelling to over 8000 members in 34 chapters spread across America, Canada, Singapore, Malaysia, Dubai, Pakistan, and, of course, India. Recognized as a key stimulus in fostering entrepreneurship in the Valley, TiE has been emulated by other ethnic groups and regularly gives advice on replicating 'Silicon Valley magic' in Europe and Asia. Since its inception, the organization has established connections with businesses worth hundreds of billions of dollars.

Most of TiE's members are first-generation immigrants who came to America in the period since 1965, when The Hart Cellar Act brought dramatic – even revolutionary – changes to America's pattern of immigration. Previously, America 'limited foreign entry by mandating extremely small quotas according to nation of origin'.[2] Under this system, very few Indians were allowed in. With the Immigration Act of 1965, however, the statutes were liberalized and quotas were lifted. The Hart Cellar Act not only significantly increased the number of immigrants allowed entry into the US, it also shifted policy away from quotas to one based on the possession of scarce skills and on family ties. As a result of this bureaucratic shift, the number of migrants from India rose substantially, from less than 2000 people in the decade of the 1950s, to close to 30,000 in the 1960s, to a quarter million in the 1980s.[3]

In the 1990s yet another threshold was crossed when, prompted by a shortage of skilled workers in the West, the American government nearly tripled 'the number of visas granted on the basis of occupational skills, from 54,000 to 140,000 annually'.[4] These visas – known by their code name H1-B[5] – coincided with the growth of the high technology sector, particularly in Silicon Valley. H1-B visas allowed companies to recruit an employee from overseas to come and work in America for a period of up to six years. Throughout the 1990s tens of thousands of H1-B visas were granted to programmers and computing engineers. Nearly half of the visa recipients came from India. In 1989, the US government approved a mere 2100 temporary work permits to Indians. A decade later, in 1999, it approved 55,000.[6]

The majority of these new immigrants worked in the IT field, and most of these set out for California in order to participate in the hi-tech Gold Rush that was taking place in the state. As Rachel Konrad writes in a special report on foreign tech-sector workers entitled *Chasing the Dream* 'the epicenter of H1-B power is in Silicon Valley, which trumped Los Angeles in the past decade and is now the state's largest Indo-American enclave. Silicon Valley's 314,819-member Indian community grew 97 per cent in the past decade. San Jose, California, has more Indians than 37 states, while Santa Clara County's Indian population has surged 231 per cent and is now the third-largest Indian county in the nation.'[7]

In a report entitled *Silicon Valley's New Immigrant Experience*, AnnaLee Saxenian wrote what has since become an oft repeated phrase: 'When local technologists claim that Silicon Valley is built on ICs they refer not to the integrated circuit but to Indian and Chinese engineers.'[8] Statistics vary, but most support this catchy slogan. Chinese and Indian immigrants accounted for 74 percent of the total Asian-born engineering workforce. In 1998 they were senior executives at one quarter of Silicon Valley's new technology businesses, and these immigrant-run companies collectively accounted for more than US$16.8 billion in sales and 58,282 jobs.[9]

Silicon Valley is said to have a 300,000 strong Indian techie community, with Indian-owned or-backed firms valued at US$40 billion.[10] *Businessworld*, an Indian magazine aimed at hyping the new economy, cites one estimate that 'a full 40 per cent of start ups are by Indians'.[11] Other studies claim that about one-third of the engineers in Silicon Valley are of Indian descent and by the late 1990s, 7 percent of the valley's high-tech firms – 774 companies – were led by Indian CEOs. 'Venture-capital fund Kleiner-Perkins Caufield & Byers, one of Silicon Valley's biggest VC firms, says 40 per cent of its portfolio consists of companies founded or managed by people of Indian origin.'[12]

The superstars of this 'golden diaspora', include Sabeer Bhatia, the founder of Hotmail, which was launched in 1996 and sold in 1997 for US$400 million to Microsoft; K.B. Chandrasekhar, founder of Exodus Communications whose fiber optic network carries 30 percent of all Internet content traffic and whose servers host such popular websites as Yahoo, Hotmail and Amazon; Vinod Dham, known as 'father of the Pentium chip' for his work at Intel; Gururaj 'Desh' Deshpande co-founder of Sycamore networks, who was listed in Forbes as being worth US$3.9 billion; Vinod Khosla, co-founder of Sun Microsystems; Rakesh Mathur, whose company, Junglee, specialized in virtual database technology and was sold for US$180 million to Amazon.com; and Kanwal Rekhi whose company Excelon made add-on boards for local area networks, and was bought by Novell in 1988. Rekhi is now a key angel investor[13] involved with over 50 start-ups in the Silicon Valley. He is president of TiE global and is widely touted as the 'godfather' of the Indian IT diaspora.

By 2003, the phenomenal success of these multimillionaires, alongside the thousands of programmers and engineers who were working at all levels of the industry, ensured that Indian IT had established itself at the core of the global IT economy. Today, their story is a familiar – if remarkable – illustration of the American Dream. Yet, the achievements of the Indian IT diaspora rest on two interrelated characteristics that were meant to disappear with the rise of modern capitalism: ethnicity and entrepreneurship.

The story of modern capitalism: the rise of the bureaucratic machine

In the story of modern capitalism, ethnicity, religion and culture were not supposed to play a role. Weber's thesis that the emergence of capitalism was tied to the specificity of Christianity applied only to the origins of the system. The reason the Protestant Ethic was able to embody the spirit of capitalism was that Protestantism created a culture whose traits had universal significance. Once this universality had entrenched itself, its particular cultural distinctiveness would fade away. Capitalism would then rest on its own mechanical foundations, no longer needing religious support.

Weber felt that this stage had already been reached by the time he was writing. 'The religious root of the modern economic outlook is dead,'[14] he proclaimed. Modern capitalism had ceased to be an expression of a spiritual commitment or religious way of life and now operated as an 'iron cage'. 'The capitalist economy of the present day', wrote Weber, 'is an immense cosmos into which the individual is born, and which presents itself...as an unalterable order of things in which he must live.'[15] Caught in this iron cage, the duties, ethics and beliefs that gave the system its cultural significance now 'prowl about in our lives like the ghost of dead religious beliefs'.[16]

Marx, of course, went further than Weber in that religion, 'the opiate of the people' had no place in his narrative from the start. Marx's aim was to create a revolutionary theory in which the only cultural difference that mattered was class. 'The history of all hitherto existing societies', he famously wrote, 'is the history of class struggle.'[17] For Marx, any other cultural, ethnic or religious distinction faded to insignificance next to this absolute divide. As capitalism developed, the two great classes – bourgeoisie and proletariat – would face each other throughout the globe as two homogenous wholes.

For these theorists of modernity, the bonds of culture, ethnicity and religion were replaced, under capitalism, by a cultural system based on universality, rationality and law. In this rationalist capitalist culture, personal, family and tribal ties no longer mattered. Instead, a system of rules would arise in which everyone would be treated the same.

The mechanism for administering these rules was considered to be the central institution of modern times. Max Weber named it the 'bureaucratic machine'. In extreme contrast to other social systems, which regulated relationships through individual privileges and bestowals of favor, bureaucracies sought instead to eradicate personal ties by operating with precision, lack of ambiguity and a 'calculability' of results. Depersonalized and automatic, it countered the fuzzy randomness of human culture with the ruthless efficiency of the industrial machine. 'The more bureaucracy is "dehumanized"', wrote Weber, 'the more completely it succeeds in eliminating from official business love, hatred, and all purely personal, irrational and emotional elements which escape calculation. This is the specific nature of bureaucracy, and it is appraised as its special virtue.'[18] The bureaucrat – or human element of the machine – had no control. Nothing but 'a single cog in an ever moving mechanism', he was simply a part of a greater whole which could neither be 'put into motion nor arrested by him' and 'which prescribes to him an essentially fixed route of march'.[19]

With its impersonal calculations and standardized rules, bureaucracies, according to Weber, are technically superior to any other form of organization. 'The fully developed bureaucratic mechanism', he wrote, 'compares with other organizations exactly as does the machine with the non-mechanical modes of production.'[20] The efficiency, order, rationality and rule of bureaucracy is, according to the economist Joseph Schumpeter, an 'inevitable complement to modern economic development'[21] and an essential aspect of the success and growth of the capitalist system.

The tale of modern capitalism, and the rise of the bureaucratic machine, like all good stories, has a beginning, a middle and an end. Eventually, so the story goes, the efficiency and order of bureaucracy subsumes all remnants of human culture and creativity. In a somewhat ironic twist to the tale, this loss signals the end of the capitalist system itself. As Weber wrote, 'in all probability someday the bureaucratization of society will encompass capitalism, and the "anarchy" of free enterprise will give way to the benefits of bureaucratic "order"'.[22] The rise of the bureaucratic machine precisely coincides – in this modernist narrative – with the triumphant dream of the socialist state. Implicit in the writings of Marx and Weber, this is stated quite openly by the economist Schumpeter. 'I for one', he writes, 'cannot visualize, in the conditions of modern society, a socialist organization in any other form than that of a huge and all embracing bureaucratic apparatus.'[23]

The process through which the capitalist system is swallowed by its own bureaucracy is twofold: First, there occurs, what Marx called, 'the concentration of capital'. This trend in which 'the larger capitals beat the smaller'[24] creates a situation where the middle tier of trade and commerce disappears and all wealth is centralized in a few hands. In this way, 'the very large, modern capitalist enterprises', which Weber says, 'are themselves unequalled models of strict bureaucratic organization',[25] begin to wipe out small and medium sized businesses that are not governed by the calculable rules of the machine. The bureaucracy of big business thus destroys what historian Braudel referred to as the 'layer of the market economy', an adaptive, creative, innovative level – below big business monopolies – on which the capitalist system continuously falls back, and from which it draws its strength.

Having ousted the small and medium sized businesses 'the bureaucratic machine' further stifles the creativity and innovation on which the system depends by routinizing the entrepreneurial function, and thus drawing capitalism nearer to its end. No one states this idea more clearly than Schumpeter in his book *Capitalism, Socialism and Democracy.*

Schumpeter is the economist most responsible for illustrating the key role that the entrepreneur plays in the capitalist system. The bourgeoisie, Schumpeter wrote, 'depends on the entrepreneur and as a class live or die with him'. The entrepreneur is the agent of socioeconomic innovation. Constantly engaged in creating 'new combinations', he or she is the abstract 'social function of innovation', and cannot be defined either as a specific person or member of a distinct class. 'The fundamental impulse that sets and keeps the capitalist engine in motion', wrote Schumpeter, 'comes from the new consumer goods, the new methods of production or transportation, the new markets, the new forms of industrial organization that capitalist enterprise creates.'[26]

It is precisely this activity that the growth of modern capitalism, 'by its very achievements', will eventually, inevitably destroy. 'The perfectly bureaucratized giant industrial unit not only ousts the small or medium sized firm,' warned Schumpeter, 'in the end it also ousts the entrepreneur.'[27] As the economic process becomes depersonalized and automatized 'bureau and committee work tends to replace individual action',[28] and 'rationalized and specialized office work will eventually blot out personality'.[29] In this situation, 'innovation itself is reduced to routine'[30] and the entrepreneur becomes 'just another office worker – and one who is not always difficult to replace'.[31]

Global tribes

In 1992, the same year TiE was formed, Joel Kotkin published a book entitled *Tribes: How Race, Religion and Identity Determine Success in the New Global Economy*. Kotkin's basic argument was that, to understand contemporary globalization, one must focus on the critical importance of networks such as TiE.

The book begins by recalling the vision of modernity based on the triumph of a 'rational and universal world order', where scientific progress and a Westernized culture superseded outdated societies rooted in ethnic and religious traditions.

Within the first few pages, however, Kotkin notes that, by the end of the 20th century, with the 'excesses of Islamic fundamentalism, irredentist chaos within the former Soviet bloc, and racial strife in American cities',[32] it had become abundantly clear that this vision had failed to materialize.

Theorists of contemporary capitalism have responded to this new reality by concentrating with a renewed interest on the problem of culture. Yet, as Kotkin points out, their focus on the rise of religious fundamentalism, ethnic conflict and tribal wars has ensured that cultural resurgence is most often viewed as an antagonistic, retrograde force, a 'throwback to the basest kind of clannishness',[33] that is seen to be inherently antithetical to modernity.

Kotkin's great insight comes from his ability to break free of the dialectical struggle, which pits Westernized modernity against the backlash of traditional cultures.[34] He envisages, instead – as McLuhan did before him – a process of retribalization, which cannot simply be viewed as an inherently oppositional trend. 'Beyond such visions', writes Kotkin, 'lies the emergence of another kind of tribalism, one forged by dispersed ethnic groups.'[35]

Kotkin is here referring to a type of vibrant economic activity that can be radically contrasted to both the theory and practice of modern capitalism. Though widely ignored, these marginal forms of trade operated throughout modernity, weaving their own subterranean tale.

In this alternative story, universalism did not take hold. Instead, race, ethnicity, customs and culture functioned to exclude certain groups from fully integrating into the mainstream. These groups (whose archetype was the Jew), consisted primarily of marginal trading people, diasporic populations and 'middlemen minorities' who – as relative outsiders – operated at the edges of society. Fulfilling their roles as peddlers, merchants and small traders, they were most at home in the chaos of markets and bazaars, where people and goods mix freely. These zones have always escaped the rigidity and order of the bureaucratic machine.

The progressive narrative of modern capitalism accounted for the economic activity of these ethnic, diasporic or 'pariah' populations in two ways. Either it was claimed that they belonged to a 'traditional capitalism', from which the modern, industrial, bureaucratic form emerged and superseded. Or, alternatively, it was believed that they existed alongside modern capitalism but were able to flourish only in backward regions 'where modern capitalism and its rational bureaucratic forms of organization and law were not yet developed'.[36] In either case, the marginal economic activity of these ethnic groups was relegated to 'the doomed periphery of the world economy'.[37]

In time – as capitalism expanded – it was thought that universalism would replace particularism and that 'these businesses would modernize and lose their distinct ethnic characteristics'.[38] The periphery would be enveloped by the core, as the 'big, rationally organized corporations displaced small and medium sized businesses that operated with traditional rules'.[39]

Yet, the idea that the machinery of modern capitalism would subsume all other 'marginal' forms of economic activity was clearly wrong. Anyone who has visited or lives in large urban centers is familiar with these marginal sectors where ethnic capitalism continues to flourish. In Western cities like Toronto, New York or London, this is evident in the dynamism of Chinatown, the ubiquity of Asian-run corner stores, which spread to almost every neighborhood, and in the vibrancy of enclaves such as little India and little Italy. Throughout the world, ethnic economies such as these have prospered and in some cases grown stronger. Modern times have seen them thrive not only within the core but also – and perhaps more importantly – in the newly developing periphery.

Kotkin calls these dispersed ethnic groups 'global tribes' and his book is dedicated to showing how they have always played an integral part in both capitalism and modernity. By combining 'what liberals had thought intrinsically separate, ethnic identity and cosmopolitan adaptability',[40] these global tribes are participating in the production of a global culture that is based on diversity, openness, invention and trade. They are, according to Kotkin, 'today's quintessential cosmopolitans'. Far from being hostile to the planet-wide technological economy, global tribes are 'the privileged actors in the post cold war globalized world'. It is likely, writes Kotkin that 'such dispersed people – and their worldwide business and cultural networks – will increasingly shape the economic destiny of mankind'.[41]

Within the past thirty years such researchers as Ivan Light, Alejandro Portes and Edna Bonacich, have developed a discourse which seeks to define the characteristics of ethnic economies and immigrant entrepreneurs. Using examples like the Miami Cuban economy, and the Korean shop-owners in Los Angeles, they have sought to show – in contrast to the theories of modern capitalism – how ethnicity plays a role in economic life.

The main aim of these theorists is to detail the ways in which ethnic economies operate differently from 'mainstream capitalism'. One of the primary distinctions is that the two economies are modeled after different conceptions of the machine. Ethnic economies have nothing in common with the bureaucratic, legalistic, impersonal mechanisms that form the basis of modern industrial capitalism. Instead, they are structured according to the principles of an information network.

In ethnic economies, shared values, beliefs, customs and institutions operate as a positive economic force. Together these create personal culturally based networks, which are used to get jobs, set up shops and finance business. Central to these networks is the concept of 'ethnic-based trust', in which reputation and common mores substitute for legalistic contracts and bureaucratic regulation.

Through their reliance on trust and social contacts, their entrepreneuri-alism and their promotion of cultural ties, the information networks of ethnic economies provide a positive alternative to the impersonal and legalistic structures of modern capitalism's bureaucratic machines.

The power of these ethnic networks is most dramatically felt in the financial sector where immigrants and small businesses have a notorious disadvantage in dealing with mainstream banks. Network-based econo-mies have developed ways to bypass these bureaucratic structures. First, by relying on loans from family and friends but also – more importantly – by devising, as a group, means of financing one another. As Ivan Light writes, 'ethnic groups provide their members financial capital through personal loans, rotating savings and credit associations, loan societies, and other cooperative endeavors, all of which rely on reputation and enduring relationships as collateral'.[42]

These network-based ethnic economies have shown themselves to be highly entrepreneurial in nature. Research shows that in America, for example, there is a dramatically higher level of self-employment among immigrant minorities than among the white majority. Where 'native white workers have resigned themselves to salaried and wage employment in the monopoly and state sectors',[43] immigrants still follow the 'American Dream' and are willing to take the risks and devote the time and energy necessary for creating new businesses and industries. To quote Kanwal Rekhi, 'immigrants by and large are footloose. They have nothing to fall back on. They are risk takers to start with. They came 10,000 miles from home and they are very mobile. They are not rooted so they can quickly move where the opportunities are.'[44]

This entrepreneurialism is a collective phenomenon. Rather than accept-ing the glass ceilings and prejudice in the work place, ethnic economies seek to turn their community's status of disadvantaged outsiders into an advantage. They do this by creating an entrepreneurial culture, which is not based on individualistic thinking or acting independently, but is instead rooted in webs of ethnic resources that are designed to help start-ups and run small businesses.

The entrepreneurial activity of these ethnic networks has fundamentally altered the predicted course of modern 'mainstream' capitalism. The bureaucratic concentration that was meant to dominate the core has been offset by a steady influx of small businesses and entrepreneurial activity. Due to the economic power of immigrant communities, the innovative and adaptive layer of the 'market economy' has not been subsumed.

Related to the idea of ethnic economies is the notion of diasporic capitalism, another concept used to describe economic activity that is marginal to the modern mainstream. Yet, despite this commonality, these two terms imply a different emphasis, and operate with their own distinct unit of analysis. Instead of focusing on immigrant minorities who settle within the core, the study of diasporic capitalism emerges out of a history of 'middlemen minorities' with a 'sojourning orientation' and concentrates instead on dispersed ethnic groups. The key notion is that of the diaspora, which is 'used to express a whole scattered group and their links'.[45] Whereas the networks of ethnic economies are closed systems that operate as enclaves within the core, the networks of diasporic capitalism are planet-wide and operate transnationally. Diasporic capitalism consists of deterritorialized populations who use host societies as a hub or node in their transnational webs of communication, finance and trade.

Historically, no people have been identified with diasporic capitalism more than the Jews. In the contemporary world economy, however, the prime example is undoubtedly the Chinese. There are approximately 60 million Chinese living outside the mainland, making the Chinese diaspora the second largest in the world.[46] The Chinese have a long history as middleman traders and a vibrant merchant culture has always prospered on the edges of the Middle Kingdom. These edges produced a dynamic maritime culture with a thriving coastal trade linking mainland China to Siberia, Korea, Japan, Indonesia, the Indian ocean, the Malay peninsula, and across to the Indian subcontinent.

Today the Chinese diaspora has spread throughout the world, creating a web of trade and commerce which has proven itself to be phenomenally successful, particularly in East Asia. In 1991 ethnic Chinese accounted for 3.5 percent of Indonesia's population but held 73 percent of the country's share of listed equity. In Malaysia the ratio was 29 percent of the population to 61 percent of the listed equity; in the Philippines 2 percent of the population made up 50 percent of the equity; and in Thailand 10 percent of the population held 81 percent of the country's listed equity.[47] These statistics make clear that the network of Chinese diasporic capitalism played an enormous role in contemporary capitalism's most dramatic story of economic growth: the rise of the Asian Tigers.

In their book, *The Chinese Diaspora and Mainland China: An Emerging Economic Synergy*, Constance Lever-Tracy, David Ip and Noel Tracy detail the specific mode of operating and particular business practices that characterize Chinese diasporic capitalism. These traits include relatively limited government support due to state indifference or hostility, a consequent lack of bureaucracy, a tendency to operate in small often family-based businesses with no separation between ownership and control, and a strategy of growth through diversification. But the most singular and important trait, distinctive to diasporic capitalism more widely, is a tendency to rely on 'trust based personal or community networks, rather than on bureaucratic structures or legally sanctified contracts'.[48]

This network of personal relationships is described by the Chinese word 'guanxi'. According to Ivan Light, 'guanxi is the ability to build useful social relationships, to stockpile these relationships, and then to call on them for business help'.[49] Through the use of guanxi the Chinese diaspora has 'developed ways of raising credit, gathering information and entering securely into agreements without depending on states and legal systems'.[50] In short, it is guanxi which creates the cultural, ethnic webs that fuel entrepreneurship and produce immense economic growth.

Kotkin's book examines a number of diasporas including the Chinese. His section on the 'Greater India' begins by stating that Indians are the 'most recent to emerge of the modern global tribes'.[51] Yet, his analysis of this highly educated, economically prosperous and technologically sophisticated diasporic population leads him to conclude that they have the 'potential to develop into the next powerful global economic force'.[52] 'India's revival, led by its wayward sons', writes Kotkin, 'could yet shake the firmament of the coming century.'[53]

TiE

When magazine and newspaper columnists write about overseas Indians working in IT, they tend to use phrases like 'network nabobs', 'networking nirvana' and – somewhat more ominously – the 'Indian Internet Mafia'. 'The Indian programmer, high-tech professionals and entrepreneurs', writes *Fortune* magazine, 'have formed an amazing web. Indians invest in one another's companies, sit on one another's boards, and hire each other in key jobs.'[54]

By creating this tight-knit web, the Indian IT diaspora operates similar to other forms of 'marginal' capitalism. Like the Chinese, it makes use of personal cultural networks in which trust and social relations are just as, if not more, important than the 'rational laws of bureaucracy'. These networks create webs of relationships which work to mobilize ethnic-based resources, and are thus able to bypass the disadvantages, glass ceilings and prejudices that tended to stereotype Indians as 'techies' and not managers. The overseas Indians – especially in America – have created a collective entrepreneurialism, leading many of them to found their own companies or work in small start-ups founded by their peers. As venture capitalists and angel investors, they offer financial support to new companies that have difficulty accessing more mainstream institutions.

In short, they operate more as a culturally based network than as a bureaucratic machine.

The quirkiness of Californians, mixed with the raw friendliness of Americans and the ambitious drive of Indian IT entrepreneurs make TiE's annual event a great place to network. In just two days, one meets everyone from West coast wannabe gurus to Mumbai businessmen. TiE's Silicon Valley headquarters is nestled on a side street between the palm trees and mirrored buildings close to the Santa Clara hotel. On a visit a few days after the event, one is struck by the democratic feel of the place, and by the inevitable bustle of activity. The multimillionaire founders, legendary figures in the world of IT, give their time generously.

This high level of volunteerism is, according to Kailash Joshi, one of the founding members and current president of TiE, what accounts for the success of the organization.[55] 'We have in Silicon Valley,' he says leaning back in his chair, '850 professionals who claim that one of their missions in life is to be a TiE volunteer.'[56] Core members dedicate not only time and energy but also money, traveling to set up chapters, give advice and help prescribe policy all at their own expense.

Kanwal Rekhi, who counters Joshi's polished sophistication with a gruff sharpness that is both immediately impressive and a little intimidating, also sees people's spontaneous commitment as a crucial component of the organization. When asked to account for TiE's success, Rekhi responds with a laugh. 'Its strange right,' he says, 'it's a not for profit group so why do people put so much time and energy into it.'[57]

His answer, like Joshi's, appeals to the desire for cultural strength over and above any social or economic gain. 'It might sound silly to people', says Rekhi, 'but Indians have not been very proud of what they have achieved. The Indian groups divide and subdivide. TiE started out as a group that sought to transcend anything that divided you.'[58] This motivation, to produce a strong diasporic culture that is not limited by sectarian segregation, is stressed repeatedly by TiE members. It is what accounts for the organization's open and inclusive values, along with its adamant insistence not to engage with anything political or religious. The TiE mindset, explains Rekhi, is that 'anything which divides you – you should go somewhere else for that. Everything that unites you is here . . . Businesses unite, shared prosperity will unite, common quality image will unite . . .'[59]

This fusion of volunteerism and a desire to strengthen cultural webs feeds the most specifically Indian aspect of TiE, the mentoring program, which is aimed at combining the innovation of Silicon Valley culture with India's traditional guru system.

Mentors are committed entrepreneurs and professionals who have reached a point in their career where they have the time and energy to give back to the community, at which point they are invited by TiE to become charter members. The first criterion imposed on them, according to Vish Mishra, convener of TiEcon 2002, is to contribute to the organization. 'We tell them please don't come to take things away from it. You contribute, and indirectly you benefit.'[60]

Charter members lend their experience, stature, connections and relationships to TiE, while also serving as 'gurus' to young entrepreneurs who have an open invitation to approach the organization for help. Mentors act as advisors and angel investors, providing knowledge, start-up capital, 'entry points' and contacts that help open doors in the corporate world. As in the traditional guru system, mentors are people with experience who act as teachers and guides. Perhaps even more critically, mentors function as role models. To quote Kanwal Rekhi:

> The basic idea of TiE was it's hard for you to dream to be something that you haven't seen one of your type be. Bill Gates that's an American. I can't be like him. But if one of your type who looks like you and has a similar background, if he achieves success it becomes a very powerful role model. You begin to think 'Why not me too'.[61]

Like gurus, mentors must be open and available. 'You can't be a role model and not be accessible', says Rekhi. 'You have to be able to reach out.'[62] In addition to acting as inspiration, TiE's mentor system seeks to provide knowledge-sharing mechanisms so people do not have to make the same mistakes over and over again. 'You can learn from each other's experiences very quickly', explains Rekhi, who provides one hour to maybe half a dozen people a week. 'It used to be half a dozen people a day', he says, 'but that became untenable. I had an open door policy but that became very hard to manage. Demand is unlimited.'[63] Before my interview, Rekhi had spent an hour with the CEO of a local company.

> Any time he gets into a situation he calls me up and spends an hour. I have been his mentor for two years. So I understand the situation in his company. When things become too complex for him and he wants perspective and to bounce off ideas I become available. I enjoy it because he is a Stanford PhD and he is very sharp and I have been able to help him.[64]

This aspect of the organization has worked to create tight cultural bonds, particularly because rivalry does not play a part in the mentoring relationship. 'There is no competition between you and me', Rekhi says of the mentor's attitude, 'because we are both participating in a larger dream here.'[65]

Though it is, in part, rooted in ethnic ties, this dream is based, first and foremost, on building a culture of entrepreneurship. From the start, the organization has had, according to Joshi, a 'laser sharp focus on its mission, the advancement of entrepreneurs'.[66]

The question of whether entrepreneurs are made or born is a complicated one. 'If you ask me,' says Mr Rekhi, 'you can't make a person into an entrepreneur. Entrepreneurship is part of your constitution, your risk profile. There are people who love to work 9–5, have their beer, watch their games and there are people who are willing to work 24 hours a day. You can't convert one to the other easily.'[67] What TiE – and the Silicon Valley culture of which it is a part – tries to do is to build an 'entrepreneurial ecosystem', an environment 'where everybody dreams of being an entrepreneur and everyone is inspired to try. That way the one or two per cent who are endowed with it will rise to the top'.[68]

TiE's first conference was held in 1994. It was conceived, according to Vish Mishra, as a festival which aimed to build the organization's identity. Called simply 'entrepreneur workshop' for lack of a better name, it sought to provide a platform for those who were familiar with the entrepreneurial process to share their experiences in an instructional manner. The conference concentrated on things like 'what does it take to write a business plan, and what does it take to go for funding'.[69] Though TiEcon has become more theatrical, the practical and educational focus of the organization has not been lost. From the beginning, insists Mishra, 'the organization has been totally focused on business building, economic improvement of the society, free enterprise, market economy, and performance based systems'.[70] Kanwal Rekhi concurs that its 'sharp focus on the economy and entrepreneurship' is TiE's bottom-line. 'That's where the action is. If you don't succeed there nothing else will matter.'[71]

TiE's aim was to transform a set of disconnected links into an integrated network. 'The idea behind TiE,' says Rekhi, 'was to bring people together so a network effect takes place.'[72] It is, agrees volunteer Mukul Goyal, 'first of all a network. Any social or economic functions are simply a byproduct of this primary activity.' Like any network TiE is built as a diverse decentered structure and is based on principles of complexity, the creative power of interaction, openness, innovation and positive feedback ('virtuous cycles' and spiral growth).[73]

Diasporas and development

Theorists tend to agree that diasporic capitalism constitutes a unique current in the global economy. It operates with a transnationalism that is distinct from both 'international state-to-state relations and multinational companies with their centralized headquarters, control and decision making processes'.[74] Chinese diasporic capitalism, for example, is a unique 'force in transnational commerce' different from both 'Western capitalism, with its finely crafted contractual system and well developed welfare state and the "companyism" characteristic of the Japanese'.[75] Overseas Chinese communities have 'largely established themselves as self regulating entities, without sponsorship or protection by the Chinese state and often by evading its law, in sharp contrast with most colonists, merchants and multinationals from Europe and America'.[76]

Theorists of modern capitalism, with their 'ethnocentric' ideas of universality, could not help but be blind to the vibrant economic activity of diasporas like the Chinese. Diasporic capitalism 'had no place in the perspectives of Dependency or World Systems theory, nor did [it] follow the prescriptions for success laid down by modernization theories or their Weberian forerunners'.[77] The success of the Chinese diaspora thus challenges the progressive story of modern capitalism with its mechanistic fantasies and assumptions about the universal triumph of rationality, bureaucracy and law, since it proves that networks based on 'personalized trust and familism' are themselves 'conducive to modernization and development'.[78]

The idea of 'marginal capitalisms', which sees economic growth as emerging out of culturally based networks, is pitted against the narrative of modern mainstream capitalism with its hierarchized, bureaucratic machines in which people function as cogs. These two economic currents are not only opposed; they are in direct competition. The original presumption was that the machines of modern capitalism would eventually swallow the networks of marginal capitalism. Yet, there is now a sense that the battle is swinging the other way. In the globalized information economy, the marginal has the potential to overtake the mainstream.

In their writings on the Chinese diaspora, Lever-Tracy, Ip and Tracy argue that the standardized highly bureaucratic multinational companies have found it difficult to adapt to the increasing unpredictability and lack of control that characterizes the post-Cold War information economy. 'It is a world in which uncontrollable choices of individual actors can have unpredictable consequences,' they write. 'Yet, planning and organization were the hallmarks of the great corporations and predictability was a prerequisite for the Fordist mass production on which many of them rose to power.'[79]

Diasporic capitalism, on the other hand, with its long history of trading through transnational networks, has long been familiar with the 'disorientating experience of globalization'. An integrated world economy, fused together by information technology, only makes their situation easier. 'Diaspora capitalists are accustomed to insecurity and to operating in unfamiliar environments, relying on family management of transnationally networked small and medium sized firms, on personal trust and reputation and on strategies of flexible diversification rather than on bureaucratic structures, law and state support.'[80]

This superior adaptability is most clear in the case of developing economies, especially China. When it comes to taking advantage of the development in the Chinese mainland, the networks of diasporic capitalism have been much more successful than the highly bureaucratic MNCs. 'The unplanned and unpredictable horizontal and vertical pathways of guanxi weave a multiplicity of random connections through the whole system, which nonbureaucratic decision makers are able to seize upon. This is proving more fruitful than the predictable rationalism and universalism postulated by Weber,' they write.[81]

> Diaspora Chinese capitalism offers a sharp contrast to the established Western multinationals and even the Japanese, who have shown little ability to develop the third world in a sustained way, and little will to respond to the challenge of opening borders of previously autarkic Communist states. A capitalism that does not expand must decline and be overtaken by others.[82]

Lever-Tracy, Ip and Tracy make clear that, in their opinion, the case of the Chinese diaspora is unique. 'Through chance and timing, the vast resources of the overseas Chinese have been able to make use of huge underused resources and markets on the mainland, meshing with dynamic forces of development there.'[83] The authors explicitly compare the Chinese diaspora with the case of Indians overseas. They argue that the Indian diaspora cannot be seen as a comparable force since it is much smaller, has no capital city like Hong Kong or Singapore, and the opportunities in India are not – at least yet – as great. Yet, there are increasing signs that this difference will collapse with time and that overseas Indians will eventually play as important a role in India as overseas Chinese did in China.

For decades now India's diasporic population has served to inspire their national counterparts. Magazines in India report that, while the country's GNP was approximately US$265 billion, the Indian diaspora generated US$125 billion. 'That's correct,' writes the lively and proselytizing *Businessworld* magazine, 'the gross personal income of Indians abroad is just under half of the gross national income of India'.[84] This tremendous and visible achievement has not only raised the confidence of Indian entrepreneurs, it has also shown, both to India and the world, that 'the Hindu rate of growth' had nothing to do with Hindu culture *per se*.

The accomplishments of the Indian diaspora were – and remain – a direct challenge to the protectionist tendencies of the Indian state. *Businessworld* delights in taunting the government by drawing attention to the success of the NRIs (Non-Resident Indians).

> What happens when you take a fish out of water? It flops about helplessly, gasps for breath, shudders and dies in a few minutes. What happens when you take an Indian out of India? He takes a deep grateful breath (after all that polluted air back home), rushes to set up his own business and then begins raking in the money. It's happened before in Uganda, it's happened before in England, but all that is insignificant compared with what the Indian immigrant is achieving in the US, especially in Silicon Valley.[85]

'Indians do so well in the US', they conclude, 'because the US reveres entrepreneurs, makes it easy for them to set up business and helps them once they are on their way. In India, on the other hand, an entrepreneur has a hundred forms to fill and tens of *babus* to satisfy... Remove obstacles and Indian entrepreneurs will thrive in their own homeland.'[86]

Beyond acting as an inspiration and battle cry, the Indian diaspora has had a more direct impact in India, particularly on the growth of the country's software industry. As Indian IT workers became more established in America, they began to set up offshore development centers back home. Many Indians working in America retained their connections with their country of origin and used Silicon Valley in particular as a hub for linking transnational webs. Like their Chinese counterparts, the Indian IT diaspora makes use of their ties within the Indian subcontinent to gain access to cheap labor and business opportunities. In a study by AnnaLee Saxenian, 46 percent of foreign-born respondents reported helping to arrange business contracts back home. One especially prominent example is Kailash Joshi who was responsible for IBM's reentry into India in 1992. Many other Indians working in America have been instrumental in establishing offshore projects in India. As AnnaLee Saxenian writes: 'Indians in the US have been pivotal in setting up the Indian software facilities for Oracle, Novell, Bay Networks, and other Silicon Valley companies.'[87] 'Indeed India's far-flung IT billionaires, are among the most aggressive investors eyeing Indian startups. Non-resident Indians are now showing a lot of interest in India because they think there is a lot of money to be made,' says Azim Premji, chairman of Wipro. 'They're suddenly finding that the return on investment in India is better than the return on investment in Silicon Valley.'[88]

By the turn of the millennium, Indians in America began to combine this sense of business opportunity with a social conscience as they turned with increasing interest and concern to the country they had left behind. In TiE circles, this tendency toward activism has been named social entrepreneurship. Mr Joshi explains it through a further elaboration of the familiar 'teach a man to fish story'.

> You cook fish and you feed people the fish. They then go and digest and then again they need more fish. The second way is to say 'let me show you how to fish. Then tomorrow you don't worry, they will go and figure it out. The third way is to create a pond and have fish in there... Social entrepreneurship shifts towards building the pond and teaching rather than feeding.[89]

The goal – as always – is to keep the focus on entrepreneurial activity. Central to this is a vision best laid out by C.K. Prahalad, who talks about the huge entrepreneurial opportunity that exists in serving the world's 4.5 billion poor.[90] Addressing TiEcon 2002, Venture capitalist Vinod Khosla elaborates on this theme by suggesting that social entrepreneurship will drive the future of technological growth. 'The environment, energy, distributed development all have technological solutions', he says, and 'it is entrepreneurs not big companies, that will find them.'[91]

In response to this call for social entrepreneurship, IT workers in America have begun to contribute and establish aid organizations that work in India. Prompted by the tragedy of the Gujarat Earthquake, for example, a number of IT professionals associated with TiE established the America India Foundation,[92] an umbrella organization coordinating aid groups in the US and NGOs serving India. In addition, TiEcon hosted a number of booths doing charity work, including Asha for Education,[93] an International non-profit organization dedicated to providing basic education for the underprivileged in India, and the Foundation for Excellence,[94] which supports students – especially female students – who are financially needy and have demonstrated strong academic skills.

TiE itself has begun to set up 'digital equalizer' (DE) centers in an attempt to bridge the digital divide. DE centers are located in areas 'where children have not only not touched a PC before they may not have seen one', explains Joshi, 'that is the criteria, we set up in a place where there is no hope in people's lifetime that they are going to see a computer'.[95] The centers consist of a very sophisticated room equipped with general computers hooked up to the Internet and a server. Material for grade 1 through 12, including physics, chemistry and math, in both English and the local language is provided through the server. 'We also give them an instructor and three years maintenance and we tell them don't restrict anyone from coming.'[96] The first DE center was started in a remote village in interior Karnataka on 15 August 2001, India's Independence Day. In December an all-India contest for children is held in Mumbai. The school with the DE center had 'a 7th grader, a 12 year old girl', says Joshi with beaming enthusiasm, 'she just took to this thing. She participated in this contest where there were kids from all over India, kids from the most modern schools and she took all the gold medals.'[97] Joshi estimates that the center costs about US$30,000 (much less after tax rebates) but he is certain 'that the center has paid off ten times over already just because she has brought a name to that school. Other kids are going to get encouraged. To me that is social entrepreneurship.'[98]

Examples like this have sparked an enthusiasm in the IT diaspora, and encouraged an ever-growing optimism about the country's prospects. This hope, and its contrast with the reality on the ground, has prompted an even larger dream which the Indian IT diaspora has come to call the 'movement for economic freedom'. They compare this with Gandhi's movement for political freedom. 'Just as the struggle for political freedom started in South Africa, the struggle for economic freedom starts in Silicon Valley',[99] says Kanwal Rekhi.

In the simplest terms the movement for economic freedom consists of liberating the Indian economy, and allowing the country to prosper in accordance with its potential. 'The Indians used to talk about themselves as being an elephant,' says Rekhi, 'an elephant is slow and cannot run. So I started to say No, we need to teach this elephant how to dance.'[100]

Where Gandhi's focus was on the injustices of colonialism, the focus of economic freedom is on poverty. 'I started to compare what we need to do with what happened in the 20's', explained Rekhi. 'We need to issue a clarion call to all Indians and ask: why are we so poor? The freedom movement came from Indians who went overseas and learnt liberal thinking, the rule of law, and the legal process from the British. In the same way economic freedom will come from Indians who have learnt about the efficiency and power of the free market from working in America.'[101]

'The Independence movement was spearheaded by the likes of Gandhi and Nehru, all Indians who came back from overseas. It is our turn now to liberate India' says Vinod Dham, the 'father' of the Pentium chip and CEO of Silicon Spice. Kanwal Rekhi, the grand old man of the valley takes that even further: 'We are first rate people who live in a third rate country; this has to change.'[102]

> For us to be twice as rich, we have to produce as much wealth again as we already have. The only source of new wealth is through entrepreneurship. Unless we focus on entrepreneurship we will always have this poverty. That became the economic freedom movement and that became part of the TiE process. It has caught people's imagination.[103]

Within India the movement of its most skilled population is sometimes perceived as a brain drain and most often viewed with alarm. Rekhi and others have answered this accusatory anxiety with a typical quick rebuttal, 'better a brain-drain than a brain in a drain'. Yet, the problem of the brain drain – or at least the perception that it is a problem – is beginning to turn back on itself and the question of whether there will be a reverse brain drain as happened in Taiwan and China is beginning to seem like a real possibility.

Researcher AnnaLee Saxenian suggests that the brain drain is being replaced by a 'brain circulation', and that Asians working in America are acting as 'agents of globalization by investing in their native countries'.[104]

> [They] share information about technology, jobs, and business opportunities with friends and colleagues at home. Many invest their own money in start-ups and venture funds, help arrange business contracts and advise companies or government officials in their countries. And a core group of transnational entrepreneurs have established business operations in emerging technology regions...[105]

Diasporas have played a fundamental role in Asia's recent growth, and there are signs that Indians abroad are beginning to channel their entrepreneurial energy into India's own economic development. This participation of the diaspora in the country's future promises to take India on to a path which will not be easily explained either by a modernist vision in which capitalism spreads out from the West, or by its opposing pole, the closed doors of an indigenous national development.

Diasporas bring with them a particular culture and way of doing business that is neither 'national' nor 'foreign'. Dispersed from the homeland, and yet never fully settled anywhere else, diasporas consist of deterritorialized populations that are in some way external to the protected sphere of any state. Their culture is one of outsiders, and their tradition is, by necessity, one of flexibility, tolerance, hybridity and innovation. As nomadic wanderers, they operate through planetary networks of connections, communication and commerce that are highly decentered and truly transnational. India's hopes of becoming an IT superpower crucially depends on the involvement of this diasporic resource.

5
The Technological Edge

Dataflow hits India like a snakebite! Two billion people, hungry
for success, 80% literate and willing to be educated...Cheap
satlinks and then fibes bring them together in a way they've
never known before, and opens the entire world to them.
Cultural, religious, geological, and political boundaries go down
in a few years, and the entrepreneurial spirit rises like Shiva
triumphant over the fuming corpse of the old. India is reborn,
reshaped, rivals Southern China and Korea and Russia within
ten years. India produces more software and Yox entertainment
in a month than the rest of the world does in a year...The
rupee becomes standard currency in Asia, and vies on the rim
markets with dollars and yen...

– Greg Bear [*Slant*]

MG Road in downtown Bangalore is like nowhere else in India. With its
fast food restaurants, pubs, outdoor cafes, cinemas, multistoried bookshops
and MTV-style music stores it is more globalized, more modern and
more cosmopolitan than anywhere else in the country.[1] Here, young
upwardly mobile professionals – both women and men – roam around
with their cell phones and scooters mingling with people from all over
the country, and the world. Though the sounds, smells, colors and
commotion of the place are all decidedly Indian, Bangalore is a global
city. Known as India's Silicon Plateau, it is one of the key nodes of the
information age.

Bangalore is a city built on digital technology. Software companies of
every size line the streets. Cyberpunk-style computer gray markets compete
with meticulously designed cell phone shops. Billboards are plastered
all over town advertising the latest deal in wireless. Both the downtown

core and the city's sprawl are teeming with slick call centers, high-tech research labs, world-class engineering colleges and mammoth software campuses. Bangalore is said to boast an estimated 120,000 IT professionals, over 1300 IT firms, 1000 training outfits, 600 cybercafes and 800 hardware firms.[2]

Besides this devotion to digital technology, Asia's Silicon Plateau shares many profound traits with its namesake in America. Like California's Silicon Valley, it has a large population of young, well-educated, English-speaking engineers. These inhabitants have developed an intensely entrepreneurial culture, which is continuously reinforced through vast global networks, connecting a multiplicity of small firms with large more established companies, which operate both domestically and abroad. Most importantly, Bangalore, like Silicon Valley, is truly cosmopolitan. Open and hospitable to outsiders, it eagerly welcomes the best and brightest from all over India, continuously feeding off their energy and ideas.[3]

During the industrial revolution the flows of goods and resources depended on the waterways. There was little choice but to locate the hubs of global trade at the large sea ports. The information age, however, is more virtual and, hence, more flexible. Operating through a process of soft industrialization, which need not involve massive upheavals in geography and population, the production of IT is able to occur in even the most out-of-the-way places.

The rise of Bangalore should be attributed, first of all, to the openness of its culture. But the city was also chosen for much more mundane reasons. Bangalore enjoys a near perfect climate, sunny and warm all year long without ever becoming too hot. Stunningly green, lush and tropical, it is referred to in the guidebooks as a 'garden city' and remains today – despite explosive population growth with its resultant pollution and traffic jams – a remarkably cheap and pleasant place to live.

Not so long ago, Bangalore was nothing more than a small cantonment town of little importance to anyone. It mainly attracted pensioners who considered it to be a good place to retire. Today, with over 5 million people[4], it is one of the fastest growing cities in the country, competing with Mumbai and New Delhi – India's traditional cores. A thriving center in a range of high-tech industries from software to biotech, it is Bangalore that most directly plugs the Indian subcontinent into the contemporary flows of global trade.

There is no doubt that the areas of software and IT services are driving India's socio-economic growth. Even after the postmillennial Internet bust, IT remains the fastest growing sector of the Indian economy, with an average annual growth rate nearing 50 percent. IT accounts for more than 10 percent of India's total exports. India's revenue for software services – barely US$500 million in the early 1990s – topped US$10 billion by the turn of the millennium,[5] and the industry continues to be on track to reach the Nasscom/McKinsey target of US$50 billion in export revenues and US$70–80 billion in overall revenues by 2008.[6] Most crucially the IT industry has enabled India to shed its image – both at home and abroad – of being a victimized, third-world country that is unable to compete globally. As B. Ramalinga Raju, Chairman of Satyam, one of the most successful computer companies in the country, is quoted as saying 'IT has brought an opportunity for this country to reach out to the world...[Through IT] the world has become a marketplace for India.'[7]

As digital technology becomes increasingly vital to the future of India, Bangalore, once a peripheral city, grows ever more significant. This tendency for the influence of the periphery to intensify is a key trait of the Indian IT industry as a whole, which has flourished by occupying a position at the edges of both the public and the private sphere. By retaining its strength and vibrancy while remaining on the margins, IT in India is revealing the contours of our networked future, a periphery without a core.

On the sidelines of the state

One of the key elements in the success of the software industry in India is that its virtual nature allows it to escape government control. Software development does not require massive capital investment. There is no need to set up large factories or purchase vast, expensive machinery. One of the most remarkable things about software code is that while it is becoming increasingly ubiquitous, it remains stubbornly intangible. Software technology is highly abstract. Ultimately it consists of nothing but strings of numbers, and as such it exists, first and foremost, in people's brains.

This extreme virtuality is far too futuristic for slow, archaic government bureaucracy to ever get a real grip on. Thus, for many years the Indian state was incapable of regulating software for the simple reason that it could not understand it. The advantageousness of this knowledge gap is well illustrated by a story from the National Association of Software and Services Companies (Nasscom), an industry-wide umbrella organization that lobbies the government – often successfully – to adopt IT-friendly policies. Nasscom's first real victory occurred in the historic budget of 1991 when it obtained a one-year tax exemption for software companies.

Dewang Mehta, the then dynamic president of Nasscom,[8] recalls that the organization's case rested on the argument that the import duty would be impossible to implement since software can be downloaded through Internet gateways. When they explained this to the finance minister, however, 'an officer immediately piped in: "But sir, we can make a customs officer sit at their gateway and charge the duty!"'[9] Luckily Chidambaram, the finance minister at the time, understood the difference between a gate and a gateway and the one-year waiver was granted.

This lack of government intervention has been widely welcomed by the libertarian culture of the Indian IT industry. Everyone from the young entrepreneurs of Bangalore to the most established industry professionals operate with a kind of do-it-yourself hacker ethic. As Ajit Balakrishnan, Chairman, Founder and CEO of Rediff online says, 'we are terrified of the government's attention. When I am invited to attend government committees, I tell them, please do not look our way.'[10] When the government finally got around to forming an IT ministry in October 1999 the industry's first reaction was, according to Madanmohan Rao 'Oh my God, Please, bless us with benign neglect. We are doing just fine. We don't need any ministry. Just stay out of our way.'[11]

To be fair, the government has in fact made some helpful moves, the most important of which was to institutionalize this 'outsideness' of the IT industry. They did this primarily by establishing the Software Technology Parks of India, or STPI. The STPI system was set up in 1990 to help promote software exports in the country. Undoubtedly one of the most successful government initiatives, it has grown exponentially and now has offices in over 30 cities around the country.

As an industry-wide organization it fulfills two basic roles. The first is to operate as something like a deterritorialized special economic zone. For the more than 7000 companies that operate under its umbrella, the STPI functions as 'a single window' simplifying and rationalizing the notoriously cumbersome government procedures. STPI is able to cut through red tape and bypass the normal protocols and regulations. As Manas Patnaik, head of the STPI in Bhubaneshwar explained, 'my job is to look the other way'.[12]

In addition, STPI offers a number of advantages that are not available to any other business sector. Chief among these are various tax incentives including exemption of corporate income tax for a block of 5 years within the first 8 years of operation, and the abilities both to import foreign goods and to purchase domestically without paying any duty. With STPI's help, setting up a business – normally a daunting process in

India, which takes up an inordinate amount of time and energy – becomes almost effortless. It is so easy, in fact, that at least one branch of STPI promised during an interview that, were it so desired, even I, a foreigner, could set up a software company in a single afternoon.

Besides slicing through bureaucracy, the STPI also provides infrastructure, allowing IT companies to bypass India's crumbling public sphere. One of the most dramatic contrasts in contemporary India is that between the 'vibrant private space', as Gurcharan Das puts it 'and the callous public space'.[13] This striking distinction is apparent even at the smallest social scale. Walking down the street, one cannot fail to notice the discrepancy between people's personal space, which tends to be extremely clean, with crisply ironed clothes, and spotlessly polished shoes, and the filthy, unpaved, garbage-littered public areas within which they are forced to walk.

The starkness of this contrast, writes Das, causes great anxiety among the local population. It is also quite disconcerting to foreigners. One international businessman speaks of this disparity as he tells of visiting his Indian partner's home. Led through a run down alleyway and into a decrepit building, he became increasingly concerned. He began to question his previous assumption that his partner was a successful businessman. Yet, upon arriving he was both surprised and relieved when the door opened into one of the most beautiful apartments he had ever seen.

This dualism reaches almost comic heights in Mumbai where arcane rent-control laws allow many long time residents to live in downtown apartments, where real estate is among the most expensive in the world, practically rent free. This situation ensures that neither landlord nor tenant is motivated to invest in public spaces, making any urban development particularly difficult. This is a great tragedy for Mumbai – India's finest city – as much of its prime property remains shockingly derelict. Walking into the office of a successful travel agent in central Mumbai, for example, requires entering a structure so neglected that it is forced to post a warning sign that reads 'Enter at Your Own Risk'. Once past the rickety staircase, however, one finds that the building is filled with the clean, modern, fully wired offices of highly successful businesses.

This contrast between the public and the private sphere is also felt at the most macroeconomic level and is illustrated most harshly by the nation's power sector. Almost anyone who relies on the public electricity grid in India suffers daily power cuts.[14] To counter this, middle-class families are resorting to buying their own generators or inverters. Thus, in modern-day India, a private back-up power supply has become – along with a cell phone and TV – a typical middle-class purchase.

For IT companies, who require highly efficient transportation and communication grids in order for their businesses to function, it is imperative to escape these failures of the State. The government, in an implicit recognition of its own ineptitude, helps facilitate this process. The STPIs, which are in the business of offering high speed communication links, also provide centers equipped with back-up power, telecom services, bandwidth and other technical infrastructure. These centers or offices are offered as incubation for small and medium sized companies who are able to use these services while they search for a more permanent base.

On a much more ambitious scale, the Indian government, most often in partnership with private enterprise, has set up a number of science parks. It is here, in these vast cybertownships that the extraterritoriality of the IT industry is most obvious. These newly constructed cyber-cities are equipped with state-of-the-art facilities including 'reliable power supply, drinking water and even transportation'. They serve as self-contained 'gated business communities' with their own restaurants, hotels, banks, recreation centers and apartment blocks. Cut off from the problems and chaos of their immediate surroundings they serve to comfort foreign visitors and 'remove everyday obstacles from the paths of their native Indian employees'.[15]

The Indian IT industry, then, has thrived by finding ways to elude the breakdowns of the public sphere. Indeed, this is considered to be such a crucial requirement that the State itself has developed ways to secure this political eccentricity of the IT industry. Yet, while Indian software has benefited from its position on the sidelines of the Indian state, it has also thrived by positioning itself on the margins of the Western core. The Indian IT revolution has not occurred by being subsumed by the center, but rather by carving for itself a place at the edges, which has allowed it to respond to the gaps, tendencies and accidents that have dominated the development of global cyberculture.

The globalization of the software industry

In the early years of the computer industry, software and hardware were bundled together as a single package. Software – no doubt due to its intangibility – was viewed as a secondary or peripheral product and was given away for free. As early as the 1960s, however, software development and maintenance costs were beginning to exceed hardware costs. It was thus becoming obvious that computer hardware was, in many ways, secondary to the software that allowed it to run. In the late 1960s IBM became the first company to realize that software could be a revenue stream in itself. It, therefore, unbundled hardware from software, and began to market and sell each separately. The software industry was born.

It took less then a decade for the industry to globalize. Already, by the 1970s, it was clear that in the West, the growing area of software development was faced with a scarcity in labor supply. As the industry grew there were simply not enough trained programmers to meet the growing demand. In 1992 this gap between the supply of and demand for programmers in the United States was estimated at 37,000. More recent estimates put the figure at 190,000, while 'world-wide there are over 900,000 programming jobs waiting to be filled'.[16] Faced with this severe shortage of software skills in developed countries the software industry began to turn its attention to engineers in developing nations, who would soon prove themselves to have higher productivity, create software of better quality, and, probably most importantly, were five to ten times cheaper than Western engineers.[17]

The key to India's role in the global IT industry has always been with the country's biggest resource – its population. Indians' affinity with IT has been attributed to a number of factors including the popular propensity toward never-ending argument, the logical structure of the Sanskrit language and a deep history of mathematics. The advantages of these ancient cultural traits were further reinforced by the greatest and most long-lasting legacy of the British Raj – the English language. Independent India also made a fundamental contribution by creating some of the best engineering colleges in the world.

The most famous of these are the Indian Institutes of Technology (IITs). Established by Nehru's government in the years immediately following independence,[18] the IITs were given a high degree of autonomy and lots of money, and this undoubtedly contributed to their success.

You would not know it from looking – the campuses are pleasant but not particularly impressive – but the IITs are one of the most successful educational establishments on the planet. When the television show *60 Minutes* did a report on them it called the IITs 'the most famous university that you have never heard of'.[19] There is no doubt that the IIT alumni have had a disproportionate influence on the high-tech industry both in India and the US. 'Two IIT alumni are aides to two of the richest people in the world, Bill Gates and Warren Buffet.'[20] Rajat Gupta, head of McKinsey, graduated from IIT, as did Narayana Murthy, founder of Infosys, and Vinod Khosla, co-founder of Sun, and partner at

Kleiner and Perkins. As Ambassador Robert Blackwill said at the celebration of IIT's 50th anniversary, 'we can think of India as a technological force in the world. That vision is owed greatly to the contribution IIT made...you (the IITians) could be the hottest export India has ever produced.'[21]

The success of IIT is generally attributed to an intensely difficult entrance procedure which admits less than 2 percent of applicants – a rate that makes even American Ivy Leagues seem easy by comparison. This ability to choose only the cream of the crop is, according to Professor Anand Patwardhan who teaches Technology, Change and Innovation at IIT Mumbai, further augmented by 'a fairly unique environment and culture that has evolved throughout the IIT's'.[22] This culture involves students from across the country coming together in a relatively isolated community that is highly exclusive and intensely competitive, but at the same time open and free.

Though the IITs are the most famous institutions, India has a host of other engineering colleges that are extremely innovative and successful. Many of these have synthetic arrangements with the private sphere. The oldest and most well established is NIIT[23] which, with over 2 million students, is one of the largest IT education and training organizations in the world. NIIT, a pioneer in computer education and training, combines its learning centers with a booming software and IT solutions company operating out of India with business in 38 countries worldwide.

More recent additions are the International Institute of Information Technology (IIITs),[24] which have combined first class education with the specific industry demands of multinational companies. They are, thus, able to bring together, as their website states, 'the freedom of academia with the efficiency of the corporate sector'. IIIT now have brand new campuses in Bangalore and Hyderabad. These large, well-established institutions compete with a multitude of other training centers and schools. In almost every village and town, a host of small businesses offer courses in basic programming and web design. Together these educational and polytechnical institutions produce 'a growing resource of about 4.5 million technical workers [and] more than 70,000 software professionals every year'.[25] This makes India home to the second largest pool of English language engineers in the world. It was apparent early on that this vast pool of highly educated, highly skilled IT professionals would be required worldwide.

The Indian IT industry was quick to realize this, and – breaking with the protectionist attitudes that surrounded them – pushed hard from the beginning to enter the global sphere. In the early 1970s Indian companies, intent on capitalizing on the labor gap in the West, began to send their technical staff to work on particular projects overseas – usually to America. This practice came to be known as 'body-shopping'.

Body-shopping was an early attempt at globalizing IT labor. It consisted of filling specific openings by hiring programmers – or 'bodies' – on a contract basis to come to work on-site for a limited period of time. The programmers were generally given temporary work visas, and when their contract was up they were either assigned elsewhere, on a new contract, or returned to India.

This practice soon became so widespread that a whole industry grew up around it, with hosts of businesses set up to facilitate this kind of contract hiring, connecting Indian programmers with projects in America. In a report entitled *Linking up With the Global Economy: A Case Study of the Bangalore Software Industry*, Asma Lateef writes that by 1990 'over 95 per cent of Indian software companies were involved in body-shopping activities and of the 3000 programmers who were working in the software export sector, the majority were on assignment abroad'.[26]

While successful, body-shopping – as the semi-seedy undertow to its name implies – is a highly unstable system. There is no commitment from either the employer, who is only ever hiring on a short-term basis, or the employee, who is always on the look out for a better deal. Moreover, body-shopping does nothing to upgrade the skills of the programmers and, therefore, offers no possible career path.

Nevertheless, as Asma Lateef points out, many maintain that body-shopping was a 'necessary first step for the Indian IT industry'. First, because the 'exposure that Indian software technicians received through body-shopping activities helped to establish India's reputation in the United States'. Second, and even more importantly, body-shopping gave Indian software engineers access to technology and business practices that were not available at home. 'When it was difficult and expensive to import the latest hardware and software', writes Lateef, 'the opportunity to travel abroad was the only way that many of the Indian programmers and engineers were able to use and understand these advancements.'[27]

As the industry matured and globalization became more established, the reputation of Indian programmers grew. Significant advances in telecommunications technology coupled with companies' increasing willingness to internationalize led, in the 1980s, to a new phase in the Indian IT industry. Body-shopping began to be replaced with offshore development. Rather than hiring programmers to come to them, global companies began to set up export-oriented development centers close to the programmers themselves. For most companies this eventually meant India.

The first 100 percent foreign-owned, export-oriented, offshore software company to set up in India was Citibank, which in 1985 established Citicorp Overseas Ltd in the Santa Cruz Electronics Export Processing Zone (SEEPZ) in the suburbs of Mumbai. Yet, it is Texas Instruments, which set up in Bangalore in 1986, that is usually credited with the pioneering role. 'It all began with Texas Instruments', says Dewang Mehta, which 'started a revolution in offshore software development'.[28] Soon after these initial companies – Citibank, TI and HP (in 1989) – set up shop, almost all MNCs in the IT business quickly followed suit. By the turn of the millennium more than '185 Fortune 500 companies – that is two out of every five global giants – was outsourcing some of their IT requirements to India'.[29] Today, practically every software company in the world, including Microsoft, Oracle, Sun and Adobe, have established an offshore software development center in India, taking advantage of the ready availability of talented, high quality, low cost software professionals.

The Indian IT industry was not only driven by these powerful MNCs. It has also been continuously shaped by a number of local companies that contribute substantially to the transnational production of digital technology. The biggest and most well-established of these domestic enterprises is Tata Consultancy Services, or TCS.[30] TCS is part of the Tata group, a ubiquitous name in India, which consists of over 80 companies generating over US$11 billion annually. The Tatas manage the Taj Mahal, own majority stake in VSNL – India's main Internet service provider – and in a remarkable case of the Empire striking back, have partnered with UK's Tetley Tea.

TCS was founded as early as 1968 as a software export company, and played a leading role in everything from body-shopping to offshore development. TCS gained its first international client in 1971, and has since grown to be Asia's largest software and services company and the second fastest growing consultancy company in the world. It now operates in over 55 countries, with over 100 branches globally.

Another of India's traditional companies that has transformed itself into an IT powerhouse is Wipro.[31] Wipro started out as a trading company, but gained success by selling cooking oil in the early days of Indian independence. When IBM was kicked out of India in 1977, however, the company sensed an opportunity in IT. IBM left a lot of machines inside India that needed maintaining, and digital technology – despite the fate of Big Blue – was already growing fast. In the early 1980s Wipro decided to move into the computer business and began to manufacture hardware. Yet, as early as 1983 the company started to shift its concentration to software.

In these initial years, Wipro – like everyone else in the country – operated with an import substitution mindset focusing on creating Indian versions of the software sold abroad. It substituted Wipro 456 for Lotus 123, and Wiproword for Wordstar. Though the company was relatively successful with these imitation products, as soon as the economy opened up, Wipro rapidly globalized. It teamed up with overseas companies like Acer to serve the domestic market. At the same time it also started hiring out its software talent to the rest of the world. Throughout its history Wipro has put great emphasis on R&D, and is now involved in the development and design of enterprise, application, commercial and systems software. Today, it is the company with the largest market capitalization in the country, and has been rated the 7th best software services company in the world.

What distinguished IT from other industries in India, however, was that it spawned a host of first-generation companies, which belonged to a new entrepreneurial wave that was – for the first time – able to compete with old money and family ties. By far the most famous of these was Infosys. Infosys was founded in a bedroom in Pune in 1981. The legend goes that Narayana Murthy, along with 6 business partners, started the company with US$250 that they had borrowed from their wives. The first years were difficult, since Infosys was operating within the context of the license Raj. It took a year to get a phone line, and importing a US$15,000 computer required 15 trips to New Delhi to ask for permission. But, according to Mr Murthy, a former leftist who is now a passionate advocate of free enterprise, the company was given a second birth in 1991.[32]

Since that time, Infosys has grown to become India's largest publicly traded software service exporter. In 1999 it became the first Indian company to list on Nasdaq. By 2003 it was employing 'more than 14,000 people and boasted a market cap of 9.2 billion ... Thanks to stock options, it counted 87 US dollar millionaires on its payroll'.[33] Operating with a low-key style that values predictability, dependability and sustainability over flamboyant, erratic genius, Infosys' success is built on precise systems, processes and routines that monitor and measure all aspects of every project. Working in an industry renowned for its constant tumultuous change, Infosys emphasizes organization, dependability and the ability of technology to factor out surprise.[34] These fairly unstereotypic Indian traits have allowed the company always to meet its targets and – even in the downturn – to grow without fail year after year.

The Y2K wave

By the late 1990s the software industry in India was already firmly established. It crossed the US billion dollar mark in 1997. In that same year the industry was catapulted into hypergrowth when it began doubling every year. The catalyst was the computer bug known as Y2K.

As is now well known, when understood technically, Y2K stems from the fact that until quite recently computers were programmed to read only the last two digits of the year – assuming the prefix 19. This programming convention dates back to the early days of the computer industry when memory was scarce and expensive and each line of code was a precious resource. At that time a space saving protocol called for dates to be recorded with 6 digits (YY/MM/DD) instead of 8 (YYYY/MM/DD). More than half a century later this seemingly banal convention proved to have staggering consequences, since it soon become clear that, as a result of their two-digit dating system, many of the world's computers were incapable of making the magnitude jump necessary to register the year 2000. Instead of treating the stroke of midnight 31 December 1999 as the end of a unit in a linear succession, it was feared that cyberspace would take it to be the completion of a hundred year count, the pre-programmed signal for computer clocks to return to year zero (99 + 1 = 00). Incapable of recognizing the difference between the year 1900 and the year 2000,[35] cyberspace needed to be 'fixed' if it was to smoothly process the impending millennium. At its most extreme, this two-digit error in programming code threatened to shutdown planetary networks, erase large chunks of data and severely disrupt the technological systems on which contemporary civilization depends.[36]

As early as 1997 it was clear that Y2K was going to be 'the single most expensive accident of all time'[37] irrespective of what did or did not occur on the midnight of 31 December 1999. The expense of fixing this catastrophe fell largely on the developed countries of the West, where IT networks have been pervasive for decades and developments in computing have, for the most part, been laid on top of, rather than replacing, the software with the faulty code.

Yet, the costly calamity in Europe and America created an immense boom in India's burgeoning software industry. The Y2K scare created more software programming than any company could handle in-house. This was coupled with the severe shortage of software expertise in developed countries, particularly in what is seen as the routine and mundane task of programming. For these reasons, the Year 2000 project had to rely heavily on offshore work. Appealing directly to India's greatest asset, its numbers, Y2K allowed the country's IT industry to capitalize on the world-wide need for thousands of programmers capable of scanning billions of lines of code, hunting down and converting the faulty date. Indian companies ended up taking 'a lion's share of Y2K work (roughly US$2 billion worth of work from 1996 to 1999)', enabling them to widen their customer base and 'helping to cement their reputation for delivering good work, cheap'.[38]

As with the rest of the world, the period surrounding Y2K created a wild hype-induced boom in India. By 1999 it seemed that everyone in the country was somehow engaged in IT. Software companies and dotcoms were popping up everywhere. Traveling in India in 1999 it was common to meet people in tiny cybercafes, who barely knew how to turn on a computer convinced that their future lay somehow in IT. Stock prices went through the roof.[39] On the Mumbai stock exchange companies needed only to suggest an involvement with high-tech to see their share prices soar. Businesses – even those who had nothing to do with IT – quickly changed their name to incorporate the prefix 'cyber', 'data' or 'tech' in the hopes of cashing in on the mania that was gripping the nation – and the world. Ads for arranged marriages, which are one of the best indications in the country of social trends, highlighted any and all connections to digital technology. During this period, it seemed that everyone in India either wanted to become a software engineer, or at the very least, marry one.[40]

On 31 December 1999 the world watched, tense with anticipation, as Y2K made its way across the time zones. Yet, no matter where, when it was midnight at the date-line, next to nothing occurred. There were no nuclear disasters, no airplane crashes, no market collapses, no power outages, no rioting, not even a terrorist attack. In the days and weeks that followed, a sense of almost euphoric relief was mixed with a strange anger, even contempt. Y2K (or 'apocalypse not') is now widely believed to have been nothing but a hoax, a conspiracy, a myth.

For a few months the technology sector seemed unaffected. Though there was a slight immediate drop in the post-millennium market, this was initially deemed to be the result not of scepticism, but of increasing faith, the idea being that now nothing would be able to put the brakes on the 'irrational exuberance' of the new economy. In retrospect, however, it is possible to see Y2K as heralding the dramatic downturn that shaped the start of the new millennium.

Though the US slowdown took some time to reach India, by the middle of 2001, Indian software companies began to issue profit warnings, signaling that the busting of the tech bubble had reached the subcontinent.

The worst hit – predictably – were small businesses that were devoted solely to Y2K. Recognizing that they faced what must be considered the ultimate deadline, these small time entrepreneurs rested their hopes on the next big thing. The popular choices were either e-commerce, or the conversion into the Euro, which, it was thought, would generate a huge amount of work. None of these, however, ever quite materialized and many of the smaller firms that catered to the Y2K boom found it difficult to survive.

Also hard hit were the Indian software engineers in America. In an attempt to slash costs US companies dropped the prices they would pay to body-shoppers from about US$85–100 an hour to around US$55.[41] Countless engineers – especially those who were in the US on temporary visas – were put back on the bench. By 2002 the joke was that B2B no longer meant business-to-business but stood instead for back-to-Bangalore.

Yet, the downturn in India was much shorter and less severe than the one that hit the high-tech industry in America. The country's software industry continued to grow. Though its growth rate of 50 percent had shrunk in 2002 to 30–40 percent, there was no actual decline, just a slowing down.

The large Indian companies, for the most part, were not too badly affected. Large players had limited their exposure to Y2K and had maintained diverse and sophisticated portfolios, which included reen-gineering, network management and system integration, along with basic programming and rewriting code. Infosys, for example, kept booming even after the tech bubble had burst. In the 'last three months of 2002 its year on year earnings jumped 24 per cent, while sales gained 45 per cent'.[42] Most established Indian companies never posted losses – though some failed to meet their predicted targets for a quarter or two.

The only real effect was that the recruitment tap was closed for a couple of years. Yet, according to Professor Anand Patwardhan, this too had its positive side. During the 1990s boom, he explains, traditional manufac-turing and engineering companies stopped trying to recruit top students from places like IIT since they simply could not compete with the salaries or prestige offered by the IT companies. The result was a gross imbalance throughout the country. The IT sector was sucking up the entire supply of engineers, who were all doing nothing but programming. After the bust, the best engineers have once again begun to diversify, and students have realized that they should not stake everything on a single sector.

Most in the IT industry – at least most that have survived – seem to agree that the downturn was ultimately required. It forced, in their view, a necessary consolidation and maturation of the industry. IT in

India, says Ranjan Acharya and Anurag Behar, senior executives at Wipro 'has become more serious. All that frivolous activity where everyone from textile, cement, to film companies got into software has come to an end'.[43] By the spring of 2003, after a short, sharp period of readjustment, there were signs of a recovery. Recruitment was reviving and new entrepreneurs were beginning once again to test the waters. In the first economic wave of the 21st century, it appears that Asia's Silicon Plateau is faring better than the Valley.

6
Peripheral Competencies

> India is penetrating America's core...We can barely imagine
> investing a company without at least asking what their plans
> are for India. India has seeped into the marrow of the Valley.
> – Michael Moritz quoted in *Businessweek*

Returning to India in 2003 to research the state of IT in the post-Y2K
environment there was great uncertainty about what to expect.
Though the Indian software sector had not been too badly hit, it had
been weakened. Surely spirits would be dampened and the feeling of
exuberance and hype that permeated the country in 1999 would
have dissipated. Yet, it only took one or two interviews to realize that
the post-Y2K downturn in the West had had a positive effect on
India and that the country's IT sector was in the grip of another
upsurge.

The reason was that the long accepted practice of 'outsourcing', in
which businesses hire specialist companies to perform essential yet
peripheral tasks, was given an immense boost by the recessionary climate.
Not only were global companies – in a bid to save money – outsourcing
more and more of their work, outsourcing was increasingly moving
offshore.[1] This provided an immense wave of opportunity for India's IT
industry. Capitalizing once again on its peripheral status, India created
an entirely new business sector, alternatively known as business process
outsourcing (BPO), offshore outsourcing, or IT-enabled services (ITES).
Despite the challenging environment created by the war on terror, and
a depression in the world economy, this sector advanced exponentially
in the postmillennial years. By 2002 it was enjoying an annual growth
rate of 70 percent, and easily matched the headiest days of the new
economy.

Optimistic forecasts for this industry, though somewhat sobered by the high-tech bust, are, in many ways, even greater than those made during the height of the software boom. Employment in the BPO sector promises to make a much bigger impact on the country – by hiring a wider cross section of people – than software ever could. Whereas software professionals need to have science or engineering backgrounds, back offices can employ graduates from any stream. More importantly, software engineers are still predominantly male, while women usually make up at least 50 percent of a BPO company's employees. In a country said to have the largest pool of educated, unemployed women in the world, this is bound to have a cultural and economic impact whose potential is hard to overestimate. In 2002 there was a surge of hope bubbling just under the surface that ITES would do for India what electronic hardware manufacturing had already done for China.

The ever-intensifying process of globalization, combined with the growing ubiquity of IT, ensured that, by the time the tech bubble had burst in the West, offshore outsourcing, which was already well established in software development, would shift to many other spheres. An immense amount of infrastructure was built up during the boom and the dotcom crash had little effect on the actual influence of the Internet. 'The dot com bubble, that huge overinvestment in dotcom stocks,' said Thomas Friedman in an interview upon his return from a recent trip to Bangalore, 'you know what it did? It laid all this pipes, all these fiber optic cables around the world and created all this excess capacity, which made it easy – not only easy – almost cost free to transfer data from America to India.'[2] The economic fundamentals may have been flaky, but the hype surrounding the web was very real.

A number of factors enabled India, in particular, to use the Internet as a medium to provide services globally. The increasing erosion of geopolitical barriers created pressure on companies to increase speed to market and remain operational, at least virtually, 24 hours a day. This has been hugely beneficial to India, which is sufficiently far away from all the big markets – 12 hours from the US, about 5 hours from Japan and about 6 hours from Europe – to be able to take advantage of these time differences and attract business from all three. 'In this strange way', says Madanmohan Rao 'geography has been kind to us.'[3]

Another underlying dynamic was the deregulation of India's telecom industry which occurred during the 1990s, and led to substantial price cuts in communication technology. 'Telecom liberalization has meant plummeting prices in lease lines and long distance charges...In just two years, there was a 110 per cent drop in the international private lease lines with an equally dramatic fall in long distance telephonic charges.'[4] This trend continued in the new millennium. In December 2003, *The Economist* magazine wrote that 'a report by HSBC says that the cost of a one-minute telephone call from India to America and Britain has fallen by more than 80 per cent since January 2001'.[5]

The first MNCs to experiment with India as an outsourcing base in areas other than software development were GE, American Express and British Airways, all of whom set up 'captive' centers in the early to mid-1990s. These companies soon found that the same characteristics that make India strong in software make it a good place for BPO. First and foremost, India has a huge English-speaking skilled labor force that can offer high productivity at low costs. 'An IT professional with three to five years programming experience', reports *The Economist*, 'earns US$96,000 in Britain, US$75,000 in America and US$26,000 in India. At the other end of the scale, low-grade call-center jobs that, in Britain pay a salary of US$20,000 pay less than one tenth of that in India.'[6]

As early as 1998 Nasscom and McKinsey had come out with an influential report that spoke of the enormous potential for ITES in India. Yet, it was not until the downturn struck that this newly emerging field really took off. The obvious impetus for companies to engage in offshore outsourcing is the huge financial savings it brings. 'Companies typically save 40–50 per cent by shifting their costs to India.'[7] During the recession, when CEOs were faced with enormous pressures to cut costs, these savings became too attractive to ignore and many more companies were convinced to move their back offices overseas.

India worked hard to capture the flows of this outsourcing boom. The country launched an aggressive marketing campaign, which sought to lure business by promoting India as the most attractive outsourcing location. At the same time, a massive entrepreneurial streak inside India was able – with stunning speed – to create the entire infrastructure for this industry within only a couple of years.

By 2003 the electronic back office had become 'India's new sunshine sector'.[8] The country built up its credentials as a BPO center for vendors and the comfort level for the big fortune 500 companies went up. India was attracting more and more companies, while, at the same time, existing customers were transferring more and more of their business offshore. In 2003 GE, for example, employed over 12,000 people in three centers – Bangalore, Hyderabad and Gurgaon[9] – up from one center with 200 employees in 1997.[10] Indian players were also cashing in on the boom with the 'big three IT companies, Infosys, Wipro and Satyam all launching some form of BPO services'.[11] In addition to these industry giants other new specialty players also began to do remarkably well. The period of window-shopping was over. ITES had become mainstream.

The most well-known example of back office work is call centers. These centers are designed to handle credit card enquiries, billings and customer services all over the phone. As any consumer is undoubtedly aware, the use of call centers has expanded substantially in recent years. According to a study by Nasscom, by 2002 there were 300,000 call centers worldwide and by 2003 US$60 billion was spent on call center services. A large part of this growth has occurred in India. As early as 2000, Bill Clinton on a trip to Cyberabad said 'I am told that if a person calls Microsoft for help with software, there's a pretty good chance he'll find himself talking to an expert in India, rather than in Seattle.'[12]

Businessworld magazine reports that 'in five years, 336 call centers have sprung up across the country, employing over one lakh people and generating revenue of US$1.4 billlion'.[13] Interviewed in 2003, Animesh Thakur of Hero Mindmine, an HR and training company geared to the BPO sector, attested to having personally seen the industry grow heavily in the past one year. 'Big timers have increased their capacity multifold and they are still ramping up, recruitment is shooting up, people are poaching from each other, salaries are rising, the growth is very real. Just a year back in Bombay there were practically no call centers – now the last reported figure is that there are 79.'[14]

Companies set up call centers in India because it saves them money. 'India churns out two million English speaking graduates every year whose wages are 80 per cent less than their foreign counterparts...In India a call center associate is available for Rs 45 an hour while his American counterpart charges Rs 550. The rule of thumb is an annual savings of US$30,000 for every call center employee.'[15] Yet, though these dramatic savings function as the initial draw, companies stay in India because – even at these cheap prices – the quality of work is far superior. This is because in India, unlike in the West, the typical call center employee is not a part-timer and usually has a graduate or even a postgraduate degree. 'Indian teleworkers outperform Americans in similar jobs', declares *The Economist* bluntly, 'because they treat them as serious careers, and also because they are better educated than their American counterparts, who are often college dropouts.'[16]

Though the Indian industry has some retention problems, the issue of holding on to employees, is not nearly as severe as it is in the West, with the better Indian companies managing to keep attrition rates below 30 percent.[17] Though he admits that graduates from top ranking colleges will probably only stay on the job for 6 months or so before moving on to other things, Animesh Thakur believes that 'people coming from the two tier colleges or three tier colleges will look on this as a career'.[18] 'If a company is large enough', says Brian Cravalaho of ICICI OneSource, a back office and call center headquartered in Bangalore, 'it is able to offer opportunities both laterally – allowing employees to move off the call center floor into HR, marketing, accounts, or quality control – and horizontally – from team leaders to customer, operation and call center managers. This ability to open up prospects across an organization enables the industry to offer its employees a career and not just a job.'[19] 'For an ordinary graduate', says Mr Thakur, 'this is probably the best opportunity available in India today.'[20]

Call centers, however, are only one element of the back office business. Much of the industry is now focusing their energy on what they call 'nonvoice' services. These range from credit card processing, to HR, finance and accounting, website development, market research, translation, transcription, data search, integration and analysis, network consulting, management and animation. *Businessweek*, in their cover story *The Rise of India* describes this plethora of tasks by zooming in on a single back office center.

> Inside GECIS' Bangalore center Gauri Puri, a 28 year-old dentist, is studying an insurance claim for a root canal operation to see if it's covered in a certain US patient's dental plan. Two floors above, members of a 550-strong analytics team are immersed in spreadsheets filled with a boggling array of data as they devise statistical models to help GE sales staff understand the needs, strengths, and weaknesses of customers and rivals. Other staff prepares data for rival reports, write enterprise resource-planning software, and process US$35 billion worth of global invoices.[21]

ITES, thus, occupies an entire range of services, which move all the way up the value chain from the simple data-entry used in medical transcription, for example, to complex work in research and development. Abundant opportunities exist in a host of business areas including financial services, insurance, high-tech companies, retailers, remote education, automotive industries, entertainment, aerospace and pharmaceuticals companies.

The massive scale of offshore outsourcing has spawned a whole ecosystem in India, with a host of companies springing up which are themselves aimed at serving the back office industry. NeoIT, for example, sells itself as a manager and adviser for outsourcing. Operating as a vast exchange, the company uses specialized software to connect buyers and sellers both from India and abroad. The company also offers managerial services which take control of entire projects, keep an eye on speed and productivity and deal with the cultural nuances that often complicate the outsourcing relationship.[22] NeoIT was established in June 1999 by founders who looked after Asian outsourcing for Nortel. Since then it has grown into the largest advisor for outsourcing in the country, handling some of the biggest projects to come to India in recent years.

Another example is Hero Mindmine which 'offers people solutions' to the BPO sector. Hero Mindmine partners with call centers 'end to end', handling all their human resource requirements from recruitment to training, assessment and reskilling. The company also offers classes, which train new graduates for jobs as customer service agents in call centers across the country.[23] In their offices on the outskirts of central Mumbai the waiting room is packed with eager candidates undeterred by the sounds of intimidating interviews that seep in from the surrounding offices. Hero Mindmine is confident that with the ITES industry rapidly ramping up – adding 75,000 new jobs in 2003 – the demand for skills will quickly exceed supply, and the need for thousands of such schools and training colleges will continue to grow exponentially.[24]

By taking peripheral tasks and making them their core competency, companies like NeoIT and Hero Mindmine act as a microcosm to the ITES industry as a whole. 'What we tell call centers is that it makes sense for you to outsource generic training since this not your core competency', says Mr Thakur, 'a company like us that concentrates on training can always invest in keeping its material updated and keep abreast of the latest techniques'.[25] In this way the industry is able to grow laterally, creating a vast network that is spreading at a remarkable pace.

Yet, despite – or even because of – this impressive growth, the industry has already seen its first shake out. In 2000 there was a wild frenzy as hoards of small time entrepreneurs set up call centers seeking to catch the next IT-driven wave. By the first quarter of 2001 there was a cull. Most of the small time operators had no idea where to look for buyers and generate business, and many call centers stood empty. Small businessmen were losing money. It soon became apparent that ITES was really an industry for the big players. Small fly-by-night companies simply could not survive.

Clients, explains Brian Cravalaho 'need to see that you have the financial strength to stay with it. They are outsourcing whole processes and they need to be convinced that you are capable of a complete substitution.' They also need to operate on a huge scale. 'Someone like

MCI is not looking to do 100 seats out of the US, and 100 seats out of India. They want to have 5000 seats in India. That's when the cost savings really kick in. Otherwise the hassle of operating out of two offices is just too high.' Most importantly, the supposed marginal tasks of the 'back office' are actually what most customers see. What companies 'are handing out to you is much more sensitive than software – they are handing over the entire customer franchise. You are the face of this company. They, therefore, need you to have a brand name. They need to see you have as much to lose as they have. If you are too small you don't have enough to lose.'[26]

This is why, as the industry continues to grow, more and more deals are going to fewer and fewer companies. Over the past year or so, all big contracts have been won by 5 big firms: TCS, Infosys, Wipro, Satyam and HCL. This trend is being mirrored in the software sector with Satyam, Infosys, Wipro and TCS getting most of the work.[27] As the industry matures most analysts believe that this consolidating trend will continue. Today, rather than worrying about competition from a myriad of smaller companies, these larger players are together focused on a brand building exercise. 'As of 2002', explains Sunil Mehta of Nasscom, '25 large Fortune companies outsourced to India. The question is how can we change that 25 to 50? Each of these is not outsourcing more than US$50 million. How do we increase that to US$100 million?'[28]

If the Indian IT industry plays its cards right these targets should be easy to meet.[29] Theoretically, 'BPO is an almost infinite market as all companies do things that are not part of their core operations.'[30] According to a study by Gartner, organizations with revenues between US$500 million and US$3 billion currently outsource up to 30 percent of their IT work. In the coming years this figure will increase substantially. The worldwide BPO market stood at US$208 billion in 1998, Gartner reports, and was expected to grow to US$543 billion by 2004.[31] The majority of this work is still located in the West. With most outsourcing happening locally, the tendency to shift outsourcing offshore has only just begun. *The Economist*, for example, quotes a Forester report stating that '60 per cent of Fortune 1000 companies are doing nothing, or are only just beginning to investigate the potential of offshoring'.[32]

Though there are concerns inside India about bubbles and hype, all signs indicate that the country is set to win big from this increasing trend. Though it 'currently enjoys a first to market advantage and owns anywhere from 80 per cent to 95 per cent of the US offshore market',[33] India, 'still has less than 0.5 per cent – a very very small slice – of the

BPO sector'.[34] A Nasscom–McKinsey study of the Indian BPO sector predicts an opportunity of US$24 billion in 2008 – 3 percent of India's forecasted GDP. If this proves correct the industry will employ 1.5 million people directly and another 0.5 million indirectly[35] in the coming years. In a 2004 cover story *Wired* magazine's Daniel Pink cites even more optimistic figures. 'Within four years, IT outsourcing will be a $57 billion annual industry', he writes, 'responsible for 7 per cent of India's GDP and employing some 4 million people.'[36]

India's outsourcing sector faces competition primarily from the Philippines, China, Eastern Europe, Korea, Singapore, Ireland and Israel. It is likely, however, that it will be able to maintain its lead over these key competitors for at least the foreseeable future. India has both a language and labor cost advantage that will likely last for the next 10 years.[37] Countries like Israel and Ireland are already overpriced, while places like China and the Philippines still have to wrestle with the language problem. 'The offshoring business remains predominantly English speaking', states *The Economist*. This massively benefits India which has 'an annual crop of around 2 million college graduates, 80 per cent of whom speak English'.[38] When asked whether BPO is a bubble set to burst, Mr Thakur weighs his words carefully. 'There are very solid business reasons why people are coming here, cost and quality. I don't see it dying out', he concludes, 'unless we ourselves do something wrong.'[39]

Digital sweatshops?

In her article about India's call center phenomena, *The World as India*, Susan Sontag describes how 'from large floors of office buildings in Bangalore or Mumbai or New Delhi, call after call is answered by young Indians seated in rows of small booths ("Hi, this is Nancy. How may I help you?")'.[40] ICICI One Source is one such office building. Set up in April 2000, the company has since enjoyed 100 percent yearly growth and has expanded into 3 centers in less than three years. When I visited them in the spring of 2003 they were recruiting between 150 and 200 people a month.[41]

Their main office in Bangalore is reminiscent of Silicon Valley, and one is struck by the apparent casual and nonhierarchical feel to the place. In sharp contrast to the rigid, caste-driven social order of many businesses in India, here the boss jokes with the receptionist, and the people inside are young and stylish – the average age of employees is 23. Sipping a coke in the rooftop cafeteria one cannot help but feel that Indian call centers are almost cool.

This funky laid back atmosphere is offset, however, by the Orwellian methods with which the business is run. Quality, control and productivity are all surveilled through automated software, which records the time a form is opened, when it is closed, how long someone speaks on the phone, how many calls are answered per minute, and how many deals are closed. Graphs depicting the results are posted around the walls, providing a strange mixture of high-tech decoration and ominous warning.

Call center work is intensely demanding and recruitment is tough. Applicants are screened for their ability to speak English, handle numbers and deal with sometimes angry or traumatized customers.[42] It is not easy to land a job, with only one in ten people making it through the selection process.

Once recruited, candidates go through a period of intensive training that generally lasts between 4–6 weeks. Part of this consists of accent training, or 'accent neutralization' as it is called in the industry. The task of accent neutralization is done with great scientific rigor. Trainers analyze regional influences and focus on specific sounds, voice, intonation and pitch. The goal is not to acquire a British or American accent but to 'neutralize' the singsong element of Indian English that many foreigners find difficult to understand. 'It is not critical that you speak in a particular accent', explains Brian Cravalaho, 'but it is critical that you make yourself understood.'[43]

Having eliminated the distinctiveness of their accent, prospective call center workers are taught to deal with the culture of their customers. This involves everything from watching Hollywood blockbusters, to learning the history, business practice and laws of a given country, to attending seminars held by ex-patriots who share their life experience and knowledge of popular culture. In some extreme cases candidates are even sent abroad.

By the end of the training process, an employee has sufficient confidence and understanding of the cultural context of their customer that they are able to make small talk while waiting for records to appear on screen, commiserating on a local team's loss of a baseball game, for example, or sympathizing about a recent heat wave or storm – all information that is made available to the employee through the computer screens on which they work. As Sontag writes, having learnt 'basic American slang, informal idioms (including regional ones) and elementary mass culture references (television personalities and the plots and protagonists of the main sitcoms, the latest blockbuster in the multiplex, fresh baseball and basketball scores)' employees can be sure that 'if the exchange with the client in the US becomes prolonged, they will not falter with the small talk, and have the means to continue to pass for Americans'.[44]

Most call center employees are able to play-act so well – going so far as to take on a different name – that the majority of customers never suspect that they are not speaking locally. 'When her phone rings', reports *Businessworld* magazine, describing the process, 'Meghna's personality undergoes an astonishing transformation. She becomes Michelle with a faint American accent. The caller is unaware that her call is being processed by someone sitting in a Third World country thousands of miles away.'[45]

Insiders such as Brian Cravalaho insist that the requirement to act American is no big deal.[46] 'It's just like learning a new language, or putting on a new skin', he contends, 'which happens all the time in marketing and sales.'[47] Most employees approach it in a very practical way.

> People think I will do what it takes to get this thing done. When you call up a center to ask for something the last thing you need is a difficult name. You want your problem to be resolved. You don't want to have to understand a difficult name because the conversation goes topsy turvy after that. You forget what your problem was and you realize you have a new problem.[48]

Techno-theorist Madanmohan Rao supports this basically relaxed attitude. 'It is a very interesting cultural phenomenon', he says, 'to see people with Indian names who speak with Indian accents, going through a course and learning how to put on an American hat, or a UK hat, for 6 hours or so and then – when the work day ends – return to their normal life.'[49] Rao tends to view these strange cultural transmutations in a positive light, seeing them as attesting to the inherent pluralism of India. 'To pick up an accent is not that difficult', he says 'because we are exposed to so many accents all the time.'[50] In India, where people are comfortable with a multitude of identities, and are used to speaking many different languages adding 'one more hat' is not a sign of conquest but simply a further proliferation. This ability to incorporate acting speaks to the fundamental creativity of the IT industry, which is able to increase the advantages of its virtual nature – and erase the impact of distance – by playing games of pretend.

There are those, however, who see in the call center workers as sure a sign as any that Indian culture is being subjugated by the West. 'In the quest for seamless connections with their clients, call centers often give their staff American pseudonyms and train them to speak like Americans', reports *The Economist*, 'a practice that has become something of a national joke (and a badge of shame in the eyes of some commentators)'.[51]

For these commentators, call center employees are techno-coolies whose job literally depends on their acting American. Could there be, they ask, a more blatant example of economic and cultural exploitation? With its 'reputation for good stuff, cheap'[52] they accuse the Indian IT industry of helping to maintain an international division of labor in which Indians remain stuck in low-end jobs, doing routine programming and simple data entry – 'work that someone else chooses not to do'.[53] Even amongst those who feel that the industry has used 'low end work' to get in the door, and this has allowed them to slowly move up the value chain,[54] there is still concern that India merely provides cheap labor, while the agenda is set elsewhere. With 50 percent women employees, call centers are looked on from this vantage point as the sweatshops of the digital age.

Backlash

More recently, the industry has been hit by a contradictory – and ultimately more threatening – accusation. In the West, the clamor of concern over sweatshop labor has been replaced by waves of protectionist panic. Every night commentators like CNN's Lou Dobbs run scare stories with menacing titles such as 'Exporting America' in which they indict companies for their lack of patriotism and corporate greed.[55]

This growing sense of alarm has been fed by a Forrester research report which says that 3.3 million US jobs (500,000 of them in IT) will move offshore by 2015.[56]

Feeling that they have already lost blue-collar work to the third world, many in the West are desperate to protect white-collar jobs. 'In America, in Britain, in Australia, an awful thought has gripped employees in the past six months or so', writes *The Economist*. 'India may do for services what China already does for manufacturing.'[57] The uproar is growing. The cry is heard everywhere: 'India is stealing our jobs'.

This is a familiar charge, which has probably been leveled against every example of successful development outside the West. 'It all sounds so 20 years ago – when the threat to economic prosperity and national sovereignty was not Indian coders but Japanese autoworkers', writes *Wired* magazine's Daniel Pink. 'Back then, the predictions were equally alarmist – the "hollowing out" of America, people called it. And the prescriptions were equally blunt – trade sanctions and "Buy America" campaigns.'[58]

The shift of pity to fear is, in many ways, little more than a sign that – like Japan and China before it – India's economy is finally successful enough to have reached the Western radar screen. Yet, while hardly unprecedented, the anxiety produced by India has its own particular traits, and there is a growing concern that, because India deals in services rather than manufactured goods, it may have a much more far-reaching effect on America than China or Japan ever did. As *Businessweek* magazine reports, 'manufacturing, China's strength – accounts for just 14 per cent of US output and 11 per cent of jobs. India's forte is services – which make up to 60 per cent of the US economy and employ two thirds of its workers.'[59]

This figure is greatly exaggerated since, as a report by McKinsey points out, the majority of service sector jobs are in industries such as retailing, catering and personal care which can not – by their very nature – move offshore.[60] Nevertheless, alarm over the so-called jobless recoveries in Western economies has meant that protectionist demands are becoming evermore intense. In Britain trade unions threatened to strike over BT's plans to shift information service centers to India, while in the US bills have been passed that ban any outsourcing of state data-processing contracts. Since there is a tendency in democracies for even the most purportedly free trade administrations to use protectionist measures to placate special interests and garner votes, the protectionist backlash against outsourcing grew steadily stronger in the run-up to the 2004 presidential election.

The Indian outsourcing sector, worried about the consequences of this trend, is seeking to confront the challenge posed by these Western concerns. This is being done through a massive industry-wide marketing campaign that hinges on a two-pronged approach. First, Indian spokespeople – along with their Western allies who support free trade[61] – aim to gently persuade skeptics about the benefit of outsourcing and to show that free trade in services, like free trade in manufactured goods, is, to use the cyberpositive business-speak, a win-win situation.

The main reason that Western companies outsource to India is that it enhances their competitiveness.[62] By shedding 'peripheral' tasks they are able to specialize in those areas that generate more value, which makes their own economies more efficient and productive. A report by McKinsey estimates, for example, that American companies gain 12–14 cents on every dollar that moves offshore. Outsourcing benefits the American economy as a whole by providing lower costs and higher productivity.

Globalization – despite the shouts of the anti-globalization movement – is not a zero-sum game. The West has nothing to lose from India growing richer. Instead, it stands to benefit greatly from the added revenue and customer demand. As *The Economist* writes 'The British once feared the rise of America's industrial might: today, both nations are vastly wealthier than they were.'[63]

Neither is the issue of job loss as dramatic as it seems. The same McKinsey's study found that 'most workers who lose their positions find another within six months'.[64] *The Economist* reports that the 'churning' of the American job market has been going on for years and that in general it creates many more jobs than it destroys. Moreover, the jobs it creates tend to be of higher quality with higher pay.[65] If history is a guide, future gain will make up for the losses felt now. 'Throughout US history, workers have been pushed off farms, textile mills, and steel plants. In the end, the workforce has managed to move up to better-paying, higher-quality jobs.'[66] Moreover, the West is already facing a skill shortage which is being aggravated by protectionist measures such as the tendency to restrict the number of H1-B visas. This labor shortage will be aggravated in the near future due to aging populations in America, and especially in Europe. There are only two solutions to this inevitable population dynamic – immigration or offshore outsourcing.

At the same time, this industry-wide marketing campaign seeks to remind rich countries that it is deeply unfair of them to start closing their doors. Protectionism is not only wrong economically it is also profoundly inhumane. 'With countries such as China and India opening their markets for products from developed countries', says Kiran Karnik president of Nasscom 'it is time that the developed countries also give us an equal opportunity given our capabilities to provide world class products and services.'[67] *The Financial Times* is even more forthright. It is 'morally repugnant', they write, 'to cut off poor countries' sources of legitimate income and export earnings'.[68] The campaign against out-sourcing, as the famous blogger Glenn Reynolds points out, is profoundly illiberal, despite the absurd moral high-ground taken by Lou Dobbs and the like.[69] After spending time interviewing a number of Indians who work in the BPO sector Daniel Pink concludes:

> Americans, who have long celebrated the sweetness of dynamic capitalism, must get used to the concept that it works for non-Americans, too. Programming jobs have delivered a nice upper-middle-class lifestyle to the people in this room. They own apartments. They drive new cars. They surf the Internet and watch American television

and sip cappuccinos. Isn't the emergence of a vibrant middle class in an otherwise poor country a spectacular achievement, the very confirmation of the wonders of globalization – not to mention a new market for American goods and services? And if this transition pinches a little, aren't Americans being a tad hypocritical by whining about it? After all, where is it written that IT jobs somehow belong to Americans – and that any non-American who does such work is stealing the job from its rightful owner?[70]

The latest wave of the Indian IT Industry is, thus, being attacked by contradictory indictments. On the one hand, it is charged with exploiting low-end (usually female) workers who are compelled to toil in sweatshop conditions. On the other hand, these same workers are blamed for stealing the jobs of the Western middle class. Ultimately, the answer to both accusations is that offshore outsourcing is the result of an innovation in high-tech globalization – the key driving force of economic growth worldwide. 'Trying to turn back that trend', writes *The Financial Times*, 'is like attempting to preserve buggy whip-making after the advent of the Model T Ford.'[71] The choice, for both India and the world, is whether to – in Bill Clinton's words – 'embrace change and create the jobs of tomorrow or try to resist these changes and preserve the economy of yesterday'.[72]

Innovating change

Western notions of innovation generally involve developing a product and taking it to market. There is little evidence that much of this sort of innovation is taking place in India, though there are signs that, in places like Bangalore, this is beginning to change. *Businessworld* magazine, the key proponent of such optimistic pronouncements, wrote, in a 2003 article entitled *India as a Global R&D Hub*, that 'over 70 multinationals including GE, HP, and Heinz have set up R&D centers in India in the last 5 years'.[73] Two years earlier the magazine had already begun reporting that 'Bangalore as an IT center was moving into a different orbit' with 'GE Bangalore investing US$100 million by 2002 in a multidisciplinary center which will have 2,500 scientists including 500 PhDs'.[74] GE is not alone in expanding its operation in Asia's Silicon Plateau. IBM is investing US$100 million over the next 3 years into its software lab. Cisco's plans to invest US$200 million over the next three years with a long-term plan to have 5000 engineers in a huge R&D facility. Sun is planning one of the largest engineering centers in the world employing 5000 people by 2005. Motorola is investing US$40 million for a center with 1500 engineers, and Texas Instrument is soon to double its engineers. HP has set up a lab with 'the sole charter to develop products for emerging markets'. Yet, despite this hubbub of activity, 'research that would lead to paradigm shifts or Nobel prizes', *Businessworld* concedes, 'is still far far away'.[75]

This type of innovation is undoubtedly important, especially as it develops to serve the domestic market. There is a danger, however, that in concentrating solely on this type of product-driven, research-based innovation, one will miss what is already highly innovative about the Indian IT industry. 'There is no question that there has been innovation', says Professor Patwardhan, 'its just innovation in a different domain'. Rather than product innovation, 'companies like Infosys and Wipro who have perfected the art of delivery really excelled in process innovation'. This service-based process innovation has propelled Indian IT to the vanguard of what *Fortune* magazine calls 'a quiet but far-reaching revolution: the globalization of white collar work'.[76]

Addressing a packed crowd at the Nasscom India Leadership Forum in Mumbai, techno-guru Don Tapscott attempts to explain this revolution, of which so many in his audience are a part. According to Tapscott, the world is currently witnessing a massive change in corporations, which he explained by referring to a famous 1937 paper by Ronald Coase entitled *The Nature of the Firm*. Coase's paper sought to answer the basic question 'why do companies exist?' His answer was that corporations were produced because of transaction costs. Vertical integration, in which different functions were all brought together in a single company, was simply the cheapest and most efficient way of operating. The cost of doing things in the open market was just too expensive.

In recent years, however, the rise of horizontal networks, with their Internet-driven business webs, have turned 'Coase's theory on its head'.[77] Vertical integration is being dismantled, and companies with top-down structures are becoming more flat. As capitalism becomes evermore global and more high-tech this trend will only increase. 'Outsourcing', says Tapscott, 'is only the tip of the iceberg.'[78]

Running through a now familiar history, Tapscott describes how the practice of outsourcing caught on in the US in the mid-1990s when, inspired by C.K. Prahalad's and Gary Hamel's theory of core competency, companies sought to focus only on their key competitive specialties. Everything else that was strategic yet noncore could be outsourced. This included HR, finance processes, bill processing, security and payroll. 'At the core of the growth of strategic outsourcing, is the realization that a specialist could provide the same service at a lower cost with better technology.'[79] To quote Tapscott the reigning motto became 'focus on what you do best and partner to do the rest'.[80]

Around the turn of the millennium, however, a transformation occurred and 'outsourcing began to hit the very heart of businesses'.[81] Citing clothes companies that no longer make clothes and computer companies that no longer make computers, Tapscott argued that now there is really nothing that cannot be outsourced. Business webs are replacing the vertically integrated corporation. Eventually, he believes, there will be 'nothing to outsource because there will be nothing "in" in the first place'.[82] The core will itself dissolve into edges.

By being intimately involved in this much more radical transformation 'India is accelerating a sweeping reengineering of corporate America'.[83] From body-shopping to BPO, the Indian IT industry has helped pioneer a network-based business model that – like all networks – is not a vertical system controlled by a central core – but is built instead out of the flows and connections of an innovative and ever-expanding periphery.

7
The Digital Dividend

> Capitalism is essentially a tool for poor people to prosper.
> – Hernando de Soto, *Commanding Heights: The New*
> *Rules of the Game*

The dark side of the Indian IT revolution is the problem of the digital divide. The popularity of this term, along with its associated phrases – the 'digital haves and have-nots', the 'information rich and poor' – attests to a growing concern that, while a small minority of people becomes evermore cyborgian, finding new and more intimate ways to connect with advanced technologies, the majority remains startlingly unplugged. The statistics are indeed overwhelming. As of September 2002, there were approximately 605.60 million Internet users worldwide; of these 182.67 million were in North America, while only 6.31 million were in Africa, and only 5.12 million were in the Middle East.[1] Though the Asia/Pacific region as a whole is said to have 187.24 million people online, South Asia, which has a fifth of the world's population, has, at least according to some estimates, 'less than 1 per cent of the world's Internet users'.[2] Citing this statistic while addressing an audience at the Hyderabad IT forum, Ms Aruna Sundarajan, secretary of IT in the southern state of Kerala, stated that 'New York city alone has more than twice the number of Internet users than India as a whole.'[3] While the privileged few use land lines, mobile phones, satellite hook-ups and cable modems to hook into a worldwide communication network, on the other side of the divide is 'more than half of the world's population that has never made a telephone call, [and the] 19 out of 20 people that lack access to the Internet'.[4]

Nowhere is the digital divide more intensely felt than in India, a country that – at least when viewed from the prism of IT – is entering the 21st century radically fractured from within.[5] In the urban centers of New Delhi, Mumbai, Bangalore and Hyderabad there exists a highly cosmopolitan population of entrepreneurial youth who are fully integrated with the outside world and are at the forefront of developments in IT. Yet, in the small villages of the countryside, where the majority of Indians live, there are millions of rural poor who have no access to – or even knowledge of – these ever-advancing technologies. 'Two distinct India's are emerging', says Professor C.K. Prahalad, 'an enthusiastic, globally competitive India and an India of the very poor and disenfranchised.'[6] How the relationship between these two Indias unfolds will have dramatic effects on the future of the country and its place in the world.

The role IT can play in development is subject to much discussion and controversy. Inside India, there is no doubt that the IT industry has created jobs, strengthened exports and made substantial contributions to the country's economic growth. These trends are expected to continue. Nasscom, for example, has made a 'promise to the nation' that by 2008 IT will grow to be a US$77 billion industry, and contribute 7–10 percent of the national GDP. Perhaps even more importantly, however, India's IT industry has provided an example – in stark contrast to the protectionist advice of leftist theories – that India is capable of competing globally in the most 'advanced' spheres of the world economy and, to use the words of Narayana Murthy, 'add value to the global bazaar'.[7]

To the proponents of IT powered development, however, this contribution is too indirect. Instead, their aim is to bridge the digital divide as quickly as possible by making IT affordable and accessible to all. The merits of this goal have met with a certain degree of skepticism, most notably from Bill Gates who, addressing a conference on the digital dividend – a term that will be elaborated below – surprised his techie audience by saying that health care was far more important than computers. According to Gates, those living on less than a dollar a day have more pressing concerns than their lack of access to IT.

This statement is somewhat ironic, coming, as it does, from for the world's richest IT billionaire. Computers, as Gates is undoubtedly aware, are not mere luxury items. In the developing world especially, information technology has a number of benefits that are crucial to the day-to-day lives of even the most poor. One of the most important is that they facilitate access to documents and records, which were previously exceedingly difficult – if not impossible – to obtain. This is illustrated most dramatically by a project called 'Bhoomi', initiated by the state of Karnataka, home to Bangalore.[8]

The idea behind Bhoomi is quite simple. Karnataka has 7 million farmers, and 20 million land records. In the past, these documents consisted of brown, crumbling paper filled with baffling, incomprehensible scribbles. They were kept in a chaotic system where access was opaque, and governing officials had no accountability. The aim of Bhoomi was to remedy this situation by using specialized software to computerize this vast collection. Access to these land records is absolutely essential, since, without them, rural farmers cannot prove that the land they are on actually belongs to them. Bhoomi eliminates this problem by creating a method – the first of its kind in the country – to provide electronic printouts of land record certificates for Rs 15 each.[9]

Anyone who doubts the vital significance of records such as these need only look to Hernando de Soto's extremely influential work the *Mystery of Capital: Why Capitalism Succeeds in the West and Fails Everywhere Else*. Summarized briefly, de Soto's argument is that private property rights are essential to modern development. What he means by private property is not the asset itself – since many of the world's poor do actually own some kind of dwelling – but rather the representation of that asset, which allows the owner to leverage their property as capital. Access to this representation of property – through land records and titles – is the way in which capital is produced. Without it there is no way to extract the potential from one's assets and enter the system of credit and trade, the key to escaping from poverty.

In addition to providing critical documentation, digital technology has the power to make governments more accountable, since, as advocates of e-governance have shown, corruption tends to rely on face-to-face encounters. By eliminating personal meetings, IT intervention, though it need not – even should not – have this as an explicit goal,[10] tends to be one of the most effective ways of attacking corrupt bureaucracies.

Before Bhoomi, for example, obtaining land records required connections and bribes. The documents themselves were open to random changes, and the entire system was riddled with shady dealings. In contrast, Bhoomi is able to guarantee efficiency, fairness and the elimination of VIP treatment for those who can pay the biggest bribes by installing software that operates on a first come, first serve basis. It also uses farmer-friendly IT kiosks, which – in an innovation that has obvious advantages to many other spheres – ensures honesty and transparency by using a second monitor that faces outwards so it can be easily seen by the client. Obtaining land records in Karnataka is now, thanks to Bhoomi, open, transparent and fair.

Speaking in Hyderabad at the beginning of 2003, Rajeev Chawla, director of the revenue department for the Karnataka state government, claimed that 7 million people had visited Bhoomi since its inception in 2001, and that it had made 11 crore rupees in just over a year. Bhoomi now encompasses 1000 kiosks, with each kiosk holder earning approximately Rs 4000 a month. The project has been so successful that it is now expanding to include a personalized database for each farmer (including valuable information such as which land fertilizer is best suited for his particular area). The project is currently being replicated throughout the country and there is now even talk of creating an urban Bhoomi.[11]

Finally, the poor need IT because it gives them access to information, an essential resource in the contemporary economy, regardless of one's income bracket. 'Voices', the Bangalore-based NGO, found that in their village project one of the most popular uses of technology involves searching the Internet for the market prices of local goods. This information is then broadcast over a loudspeaker at the weekly bazaar. In this way, the village has been able to break long-held monopolies on information, cut out the middlemen and obtain a fair price for their goods.

This use of IT, which allows poor farmers in remote areas to gain access to market information, is one of the most important aspects in bridging the digital divide. Capitalism, as an economic system, depends upon the free flow of information, which – at least in theory – is transparent and open to all. The rural poor, however, often live in remote areas cut off from modern networks of communication technology. They, thus, have no way of knowing vital facts such as the price of their crops, the weather forecast for the upcoming season or the latest technology in farming practices. All of this has a massive effect on their ability to compete. Lacking access to market signals, the 'third-world' farmer has no choice but to rely on middlemen who take advantage of their ignorance, buy their products well below market price, and keep the farmer in a permanent state of feudal dependency.

In June 2000 the Imperial Tobacco Company of India Limited (ITCI), one of the largest exporters of agri-commodities in the country, and known by locals and visitors alike as the maker of 'Wills' cigarettes, launched a scheme to try to help farmers break out of this 'vicious cycle'.[12] The scheme is known as e-Choupal,[13] from 'choupal' which means village square. The project initially began with Soya farmers in Madhya Pradesh but has since grown to include 'around 11,000 villages across 14 states, reaching out to 12 lakh farmers through over 1,900 kiosks'.[14] The concept of e-Choupal is straightforward. The aim is to use the Internet to connect people in rural areas with the information that they need from crop prices, to facts about the soil, to the latest local and global information on weather, to tips on scientific farming practices. Kiosks are set up close to the villagers' homes and use Hindi as well as English. They are managed by a 'Choupal Sanchalak' – 'himself a lead farmer – who acts as the interface between the computer terminal and the [other] farmers'.[15] By eliminating the numerous intermediaries, the e-Choupal scheme has both increased productivity, and enabled farmers to get more for their crops. At the same time ITCI now has a direct communication link to their rural suppliers, and no longer needs to purchase crops from middlemen who add cost to the supply chain without adding value. The company thus enjoys a 'lower net cost of procurement despite offering better prices to the farmer'.

'e-Choupal', has become the largest initiative among all Internet-based interventions in rural India. There are, however, a number of projects which work on a similar basis. One of the earliest is the Warana Wired Village project, initiated by the Warana Cooperative Sugar Factory together with the central and state governments in 1998. This project connects 72 villages in the Warana region of southern Maharashtra. Much like e-Choupal, the Warana project provides farmers access to essential pieces of information such as 'the ideal time for planting and harvesting sugarcane, the current market rates of their produce, and payments made by the factories'.[16] This is especially crucial in sugarcane farming, the key crop in the region, where even a slight delay in harvesting reduces sugar quantity, thereby affecting the weight and price farmers can get for their crop. Through the Warana project, farmers are able to punch a code into the system and get details such as when the crop was planted and when it is due for harvesting. 'This gives the farmer sufficient time to organize workers to cut and transport the sugarcane.'[17] It thereby ensures that the rural poor – who are themselves entrepreneurial – no longer have to pay for their ignorance.

India's cybercafes

Projects such as Bhoomi and e-Choupal prove that the poor need – and want – digital technology just as much as they need health care and clean water. The problem is one of access. It is clear that, in India at least, this problem will never be solved by private ownership of the personal computer. No matter how much the cost of electronics comes down, high value commodities will never be affordable to the vast reaches of the lower class. If the poor are to gain access to IT, it is essential that certain assumptions and mindsets that are inherited from the West be broken. To quote Professor Patwardhan, the 'PC-centric computing and communications model of North America is simply not applicable here'.[18] In India almost everyone agrees that the solution lies not in ownership but in some form of pay per use.

The widespread confidence in this model stems from India's experience of an earlier telecom revolution that swept through the country in the 1980s and 1990s. This revolution, which led to a radical reshaping of India's communication landscape was led by one of the most famous visionaries in the world of Indian IT – Satyen 'Sam' Pitroda.

Sam Pitroda was born in a village called Titilagarh in the eastern state of Orissa. The village had no electricity, no telephones and no running water. As a young man Pitroda left the village and eventually found his way to America. After studying at the Illinois Institute of Technology in Chicago, he began to work on the development of digital electric switching systems. By the mid-1980s, this work made Pitroda into a self-made telecom millionaire. He had fulfilled the American Dream.

At this point Pitroda's attention turned to the development of India. At that time the accepted view inside India was that telephones were a luxury item. The idea, writes Gurcharan Das, was that 'a poor farmer did not need a telephone but water, literacy, and basic health'.[19] Pitroda, however, was convinced that telephone density was crucial to wealth creation.

Pitroda believed that the best strategy in State-centered India was to work to change this mindset from the top down. With this in mind, he managed to develop a relationship with Rajiv Gandhi who, as an air pilot, was, if not a techie himself, at least tech-friendly. With Rajiv Gandhi's help, Pitroda founded an organization called the Center for the Development of Telematics (C-DoT), which he ran with American-style openness and enthusiasm. C-DoT's mission was to develop a digital switching device suitable for India.

The key to Pitroda's revolution was in using this switching device to put telephones in the bazaar. 'The coin operated public phone was not a good idea', explains Das, 'because it was expensive and it quickly deteriorated in the heat and dust of India. [Pitroda] and his team came up with a simple idea: equip ordinary telephones with small meters; put these in the hands of thousands of entrepreneurs, who would set them on a table in the market.'[20]

Following Rajiv Gandhi's assassination, Pitroda lost his connection with the government and after a period of hardship returned to America. His scheme, however, created profound and lasting change. By 1998 there were half a million Public Call Offices in markets across the country, creating 1.5 million jobs. Like donut shops in Canada, or diners in America the yellow signs of STD booths are ubiquitous in India. They exist in all but the most remote market places. In 1988 only 4 percent of Indian villages had access to telephones. Today, more than half of all Indian villages have a public call center.[21] In larger urban centers one finds them at practically every intersection. Thanks to Sam Pitroda, phoning from India, which used to be a full-day affair, is now simple, efficient and getting cheaper all the time.

The C-DoT story clearly demonstrates that increasing access to communication technology requires innovation. Addressing a reverential audience at TiEcon New Delhi, Sam Pitroda spoke of the impossibility of importing digital switching for the high traffic, low density and extreme climate of India. He passionately articulated the need to develop new technology to serve these local conditions. India, he insists, is not like America. Here 'people tie water buffaloes to telephone switches, cats live in them, lizards roam around the circuitry ... India', Pitroda likes to say, 'requires Indian solutions'.[22]

Sam Pitroda's 'alternative vision', which stresses access over density,[23] applies to the Internet just as much as to telephones. Manned public call booths are what enabled many in India to make their first telephone call. In the same way, manned Internet kiosks are currently enabling millions of unplugged Indians to go online. On the periphery, cybercafes – little more than a hip accessory in the West – are key to bridging the digital divide.

These cybercafes are popping up in even the remotest villages throughout the country. Whether it is a couple of computers hooked up at the local temple, which allow kids to learn Word and Power Point, or NIIT's 'hole in the wall' experiment which introduces urban slum children to the Internet, or busses that give online access through mobile labs, 'India', as Celia Dugger for the *New York Times* writes, 'is fast becoming a laboratory for small experiments linking isolated rural pockets to the borderless world of knowledge'.[24]

Many of these experiments are being run by private enterprise. Pitroda himself, for example, along with his new company WorldTel are investing heavily, in the hope of creating a revolution with Internet kiosks 'just like the one we created for PCO's'.[25] At the same time, almost every state government in the country is trying to encourage the development of cybercafes. Andhra Pradesh, for example, is aiming to set up 5000 IT kiosks. Their strategy is to provide microfinance to the owners of existing STD booths who will use the money to wire their stalls. 'The government should give the initial spark and leave the rest to market forces',[26] says Mr J. Satyanarayana, IT secretary of the state.

Kerala, which is used to being at the forefront of development, also has an ambitious scheme to try to catch up with the more innovative IT projects that are occurring elsewhere. With the best literacy rate in the country and the highest telecom density, one would assume that Kerala would suffer less from the digital divide. Computer penetration in this state, however, is less than 1 percent.[27] Eager to remedy this situation, Kerala is now turning its attention to other more IT-friendly states, seeking to incorporate experiences from across the country. In February 2003 it launched a scheme to create a network of 9000 community 'info centers'. Based on the STD/PCO model, it uses the government as a facilitator to help establish centers which will themselves be managed by private entrepreneurs.

Local content

The creation of a rural ICT (Information and Communication Technology) infrastructure, however, does not only require affordability and access. It also demands – especially in the most remote areas – that people be made aware of IT and learn what it can do for them. As Aruna Sundarajan says, 'we need to shift focus away from supply towards creating demand'. This is why the most successful ventures like e-Choupal and the Warana Wired Village combine access with relevant content. Though these are directed exclusively at farmers there are a number of rural projects that operate on a wider scale.

TARAhaat, for example, links kiosks in rural villages to an Internet portal which is designed exclusively for their needs and desires. TARAhaat's services are accessed through a colorful graphic interface, with bright easy-to-use roll-over animations and a voice-over, which allows even illiterate villagers to access the site. 'Haat' means village bazaar, and along with offering information on topics like health-governance, welfare schemes, astrology, matrimonies and education, the site operates as an e-commerce store geared to rural consumers. It has franchised vehicles and delivery vans and has even innovated a kind of local credit card that will facilitate payment.[28] Its goal is to deliver products and services to the unserved market of rural communities from 'where an estimated 50 per cent of the national income is derived'.[29]

Another example is the Sustainable Access in Rural India, or 'SARI' project. Run by IIT-Chennai; MIT Media Lab; Berkman Center for Internet and Society, Harvard University Law School; and the I-Gyan Foundation SARI has developed 1000 connections in 350 villages in the Madurai District of the southeastern Indian State of Tamil Nadu. Like TARAhaat, it provides regional content in local languages and is geared toward entertainment, health, education, empowerment and economic development. 'While the uses of the kiosks vary from site to site', the project reports that 'some of the most popular applications are transactions with local government, computer education, and e-mail. There have also been notable successes in human and animal health, and agricultural information.'[30]

Probably the most famous project of this type, however, is Gyandoot,[31] a digital network that links the Dhar region, a poor tribal area of Madhya Pradesh, to the outside world. Initially, both the community and other politicians were extremely skeptical that scarce resources should be put into a computer network when water, electricity and food seemed far more urgent necessities. However, after extensive consultation, the village *panchayats* (local government) and the rest of the community, were convinced to support – and finance – the project. By 2000 a network of rural cybercafes was made operational. Today this network is used to lodge complaints, report thefts, access health and farming information, trade matrimonial ads and deal with essential bureaucracy from getting a caste certificate, to obtaining scholarships, to asking for pension payments. Gyandoot won the 'Stockholm Challenge award 2000' and is widely praised as a 'breakthrough in e-government' and a model to be replicated elsewhere.[32]

Serving the bottom of the pyramid

Both Gyandoot and SARI are made possible by the use of CorDECT, a wireless in local loop (WLL) technology that was developed by IIT Chennai.[33] This new technology provides cost-effective, high quality voice and data connectivity to urban and rural areas. It is uniquely able to operate with very low power conditions and no air-conditioning. Using CorDECT, most remote villages are able to set up functioning Internet kiosks for less than US$1000 each. The technology has proven so successful that it has begun to spread outside India and is now operational in Madagascar, Argentina, Brazil, Tunisia and Nigeria with field trials underway in Iran, Egypt and Yemen.

CorDECT, explains professor Anand Patwardhan, 'started with a very simple premise, how to lower the capital cost on a new telephone from Rs 30,000/Rs 40,000 to Rs 10,000'. The developers understood that, to enlarge the market, this kind of radical cost was required. What was needed was a new technology that could move to a lower point in the price performance curve. Finding the solution depended on both technological and business innovation.

The success of CorDECT shows that, as the Indian IT industry begins to focus attention on serving the domestic market, real innovations will arise. These innovations signal a new attitude to the digital divide. Rather than being paralyzed by a hopeless, guilt-ridden despair, this new innovative approach seeks instead to convert problems into opportunities. The various projects and theories which constitute this approach have coalesced under the term 'digital dividend'. Its mission as stated on the 'Digital Dividend' website[34] is to 'identify and promote business solutions to the global digital divide – sustainable models that will allow ICT-for-development to go global, creating social and economic "dividends" in poor communities around the world'.

The theorist who articulates the notion of the digital dividend most forcefully – especially as it applies to India – is C.K. Prahalad, Professor of Business Administration at the University of Michigan Business School in Ann Arbor. Prahalad's ideas, which consist of 'a bizarre mix of hardcore cutting edge business thinking and socially responsible activist initiatives',[35] start with a basic question: 'Can we convert our seemingly insurmountable problems into global opportunities, and serve the 4.5 billion people around the world that live on less than US$1,500 per year?'[36]

Prahalad insists that with the right mindset, this can be done. 'The Third World', he asserts, 'is just a state of mind.'[37] In a talk given in New Delhi entitled *India as a Source of Innovations*, Prahalad laid out a number of orthodoxies that must be broken if the 'third world' is to free itself from its stifling attitudes and beliefs. These orthodoxies include the idea that the 'distribution of wealth is more important than the creation of wealth', that 'market forces and private industry cannot be trusted to deal with the problems of the poor', and that 'inequalities can be alleviated through subsidies...50 years of state subsidies and socialist based solutions have failed', Prahalad states firmly. 'There are still 500 million people in India living on less than a dollar a day.'[38]

Rather than to continue to rely on government spending, it is far better, argues Prahalad, 'to transform the poor into a vibrant market'.[39] If the poor are ever to prosper it is imperative that the now almost unquestioned division through which 'capitalists serve the rich and the state or NGO's serve the poor'[40] be dismantled. There is a 'misconception', he writes, 'that you are exploiting the poor by selling to them'.[41] Yet, people are served best when they are seen as customers, not dependents. Rather than being considered as a problem for the State, the poor – which Prahalad calls 'the bottom of the pyramid' – would be better off if they were perceived instead as an attractive opportunity for business.[42]

Prahalad's notion of the poor as business opportunity converges with the work of Hernando de Soto, who praises the underclass for their entrepreneurship. Working in a wide range of 'microbusinesses', the world's poor have, according to de Soto's research, amassed real estate worth US$9.3 trillion dollars. This sum, writes de Soto is 'forty six times as much as all the World bank loans of the past three decades, and ninety three times as much as all development assistance from all advanced countries to the Third World in the same period'.[43] The West tends to portray the poor as destitute beggars and starving children who lie helpless on the streets waiting for charitable donations. But, to quote Hernando de Soto:

> The grimmest picture of the Third World is not the most accurate. Worse, it draws attention away from the arduous achievements of those small entrepreneurs who have triumphed over every imaginable obstacle to create the greater part of the wealth of their society. A truer image would depict a man and a woman who have painstakingly saved to construct a house for themselves and their children and who are creating enterprises where nobody imagined they could be built. I resent the characterization of such heroic entrepreneurs as contributors to the problem of global poverty. They are not the problem. They are the solution.[44]

In order to take advantage of the vast potential that exists at the bottom end of the market, it is important, explains Prahalad, to recognize certain facts about how poor people actually live. One of the most important is that 'the poor live in a very high cost environment'.[45] Without access to credit or savings,[46] poor people are unable to buy in bulk, and, even when purchasing daily items such as soap or shampoo prefer low cost micro-sachets over the more expensive bottles or bars, which demand higher up-front payment and risk being wasted or lost. Lacking any storage capacity, and without any access to cost-saving credit cards and supermarket points, they cannot take advantage of sales, and buy everything with cash on a day-to-day basis.[47] This mode of consuming – though it requires less money at any given time – is a far more expensive way of living than that practised by rich people who live in the developed world.

Even more surprisingly, the poor are often ready and willing to pay top dollar for, what in the developed world are considered to be luxury items. Throughout Asia, people in almost every sphere spend huge proportions of their salary on high-tech goods, buying a new cell phone every year, for example, or saving for months to buy a digital camera or DVD player, which they perceive as necessities. This does not only apply to the middle class. Projects like Bhoomi and e-Choupal have shown that even the rural poor have disposable incomes and purchasing power and are prepared to devote an enormous percentage – often up to 5–7 percent – of their incomes on ICT.[48] In Dharavi, an enormous shanty town on the outskirts of Mumbai, and the biggest slum in Asia, 85 percent of homes have a TV.[49]

One of the best examples of how easy it is to tap this spending power and take advantage of the welcoming attitude to new technologies is the success of GrameenPhone, a development project based on micro-financing and mobile phones that is taking place in rural Bangladesh. This region, one of the poorest in the world, 'where average incomes are less than US$200 per year, may not seem like promising territory for a mobile phone company. But GrameenPhone has shown otherwise'.[50] The project works in conjunction with Grameen bank, a microfinance institution, which 'lends money to one entrepreneur – usually a woman – to buy a mobile phone'. She then 'sells access time to neighboring villagers, who pay for calls in cash'. The scheme has 'proved immensely popular with each phone generating an average of US$100 a month'.[51] This is more than the average monthly phone bill of even a businessman or teenager in the West.

The digital dividend approach holds that MNCs, with their reach, scale and resources should, in theory, be better able to serve the poor than governments ever could. After all, 'very few, if any, governments of developing countries can deliver services in a million places at once, yet many global corporations do just that every day'.[52] Companies, who are meant to always be on the look out for new market opportunities, ought to realize that it is immensely profitable to 'do business with the 4 billion poor, who make up more than half the world's population'.[53] There are 1 billion people who live on less than a dollar a day. 'Together', writes Hart and Prahalad, 'this represents a multi trillion dollar market . . . Companies can make fortunes serving the world's poor.'[54]

Nevertheless, this immense opportunity remains 'basically invisible to the world's MNC's [who have] so far not done that well in emerging markets'.[55] Yet, large global companies ignore the vast potential of 'low income markets' at their peril. In the future, real opportunity, claims Prahalad, lies 'not in the wealthy few, nor even in the middle class but in the billions of aspiring poor who are joining the market economy for the first time'.[56] De Soto agrees. 'The constituency of capitalism', he says, 'has always been poor people that are outside the system.'[57]

Take voice-based technology for example. Companies are gearing these technological developments toward the developed market where demand is limited. Yet in the developing world, where there is an enormous illiterate or semiliterate population, applications such as voice-based e-mail – especially in a multitude of languages – could be revolutionary. 'Where is the market opportunity likely to be greater', asks Prahalad, in 'sophisticated offerings to demanding Western customers that have many other options, or basic systems in languages such as Spanish, Mandarin or Hindi that can provide an irreplaceable service for hundreds of millions of customers?'[58]

Companies stand to benefit not only directly but also from indirect gains through the advantages of network expansion. 'The value and vigor of the economic activity that will be generated when hundreds of thousands previously isolated rural communities can buy and sell from each other and from urban markets will increase disproportionately', writes Prahalad, 'to the benefit of all those who participate in it.'[59]

The bottom of the pyramid is a new market with entirely different demands that will use technology in novel ways. Companies that take up the challenge of trying to serve this market will thus inevitably find it a breeding ground for all sorts of innovation. For smaller, local businesses it is the perfect training for the global economy. Serving the poor teaches entrepreneurs how to be innovative, how to use technology and how to cut costs. 'If you can succeed at the bottom of the pyramid', says Prahalad, 'you can succeed anywhere.'[60]

At the same time, larger MNCs will 'gain long term competitive advantage from the innovation, learning, and market intelligence that comes from participation in [this] market'.[61] They may even profit more directly, since 'sustainable product innovations that are designed to serve the poor may indeed feedback to the rich'.[62] This has already occurred with pre-paid mobile cards, which were pioneered for bottom-end markets, but quickly spread to the developed world. As Prahalad points out, 'pre-paid cards appeal to the poor because they require low initial payments – approximately US$10 to get a phone number running – but they also appeal to companies because they charge higher rates for calls and don't have to deal with billing'.[63] As this example shows, poor customers can act as 'a testing ground for new technologies that serve the global market'.[64] In the end, they may be the 'springboard for the most creative use of advanced technologies'.[65]

Creative destruction

The ability of the poor to act as a crucial source of innovation – even for rich markets – is related to the fact that technology is notoriously difficult to predict. Those who do try to forecast technological trends tend to make embarrassing miscalculations. In a speech at Nasscom's industry-wide conference in Mumbai, Partha Iyengar of Gartner outlined a few of the most notorious false predictions that have occurred in the world of IT. Chief among them is IBM's 1982 assessment that 'US$100 million is way too much to pay for Microsoft', and Bill Gates' infamous statement – made in 1981 – that '640K ought to be enough for anyone'.[66] Technological prediction is impossible because, to quote Chairman of Satyam Computers, 'economic growth depends on innovation and invention. Most of what exists now did not exist in the past, and most of what will exist in the future does not exist now'.[67]

Real innovation is not a process of steady improvement, but is based instead on the discontinuous shifts which revolutionize how – and by whom – any given technology is used. One of the most important theorists of this type of true innovation is business theorist Clayton Christensen, whose work focuses on what he calls 'disruptive technology'. According to Christenson, disruptive technologies, though rare, are of primary importance to both technological and economic change. They are, he writes, 'a core microeconomic driver for macroeconomic growth, [and have] played a fundamental role as the American economy has become more efficient and productive'.[68]

Of all economists it is Joseph Schumpeter, who Thomas Friedman refers to as the defining economist of the 'globalization system' who focuses most attention on how disruptive technologies act as the motor for capitalism.

According to Schumpeter, capitalism is not a thing but a process. 'The essential point to grasp', he writes, 'is that in dealing with capitalism we are dealing with an evolutionary process. Capitalism is by nature a form or method of economic change and not only never is but never can be stationary.'[69] Schumpeter's 'evolutionism' is fundamentally different from that of Marx, whose dialectical unfolding presumes that all the elements of a system are there from the start. Instead, for Schumpeter, capitalism is intensely creative and works by constantly incorporating the new. 'The fundamental impulse that sets and keeps the capitalist engine in motion', writes Schumpeter, 'comes from the new consumer goods, the new methods of production or transportation, the new markets, the new forms of industrial organization that capitalist enterprise creates.'[70] In Schumpeter's vision of capitalism, 'innovation replaces tradition and the present – or perhaps the future – replaces the past'.[71]

For Schumpeter, these new products, markets or forms of organization do not exist along side the old. Rather, their introduction necessarily induces a period of dramatic upheaval – usually in the form of depression.

The means by which capitalism destroys the old and replaces with it what is innovative and new, Schumpeter called 'creative destruction'. To quote the most famous passage from *Capitalism, Socialism and Democracy*: 'Industrial mutation – if I may use the biological term – incessantly revolutionizes the economic structure from within, incessantly destroying the old one, incessantly creating a new one. This process of Creative Destruction is the essential fact about capitalism.'[72]

The key point for India's attempts to bridge the digital divide is that creative destruction does not begin at the core, but arrives instead from the edges, where the equilibrium state is most unformed. Only the paranoid survive, says Intel's (Schumpeterian) Andy Grove, because there is always something stirring on the margins that threatens to devastate the existing order of economic interests.

Christensen's work describes in detail how this occurs. The 'good', well managed company, he argues, finds it difficult to lead – or even withstand – the production of disruptive technology. The introduction of the PC, for example, managed to damage, and even destroy many of the mainframe giants. Insulated from the information on the ground, these large successful companies became desensitized to the emergence of their most dangerous competitors: the new products, production processes, markets, supply sources and business organizations that flow in subliminally from the periphery. 'Disruptive technologies', writes Christenson, 'have plunged many of the best companies into crises and, ultimately, failure.'

The problem is that 'well managed' companies tend to be focused on the demands of their 'best' customers – those who will pay most for the latest product. They, thus, spend the majority of their time and money researching those products that are highest in the value chain, and concentrate on satisfying their most demanding – and profitable – customers. Yet, the attempt to meet the demands of existing customers, by focusing on the high-end of the market, can prove disastrous when it comes to determining future trends. Disruptive technologies cannot be predicted or measured by market research. When Sony introduced the walkman, for example, no one was asking for handheld stereos because no one had even thought of them. A management strategy that is based on satisfying high-end markets can only result in the production of what Christensen calls 'sustaining technologies', technologies that are designed to 'improve the performance of established products'.

Disruptive technologies, on the other hand, grow most often from experimentation at the lowest end of the market where the large, 'best' companies simply are not looking. Generally 'worse' than the technologies they are designed to replace, disruptive technologies are not intended as improvements on existing products, but act instead as a means of incorporating a whole new market that previously did not exist. When they do occur in a sphere that appears to be already occupied, 'they usually perform worse at the beginning than mainstream products and are thus not of interest to high-end customers'. They also tend to 'have lower profit margins and do not have as wide a market as mainstream products'.[73]

Yet, despite these initial drawbacks, disruptive technologies create major new growth in the industries they penetrate – even as they cause traditionally entrenched firms to fail – because they allow less-skilled and less-affluent people to do things previously done only by expensive specialists in centralized, inconvenient locations. In effect, writes Christenson 'they offer consumers products and services that are cheaper, better, and more convenient than ever before'.[74] It is this which gives them the potential to disrupt and – creatively – to destroy.

According to Mukesh Ambani, chairman and managing director of Reliance, one of India's biggest high-tech firms, the Schumpeterian philosophy is not at all alien to India. In an article entitled *Preparing the Indian Mind for the New World*, Ambani writes that 'India began to decline only when it began looking inwards.' To participate and compete in the global economy requires a universal outlook which has 'creative destruction' at the core. This 'spirit of creative destruction', he writes, 'is imbibed in our epics. The trinity, Bhrama, the creator, Vishnu, the preserver and Shiva, the destroyer, epitomize this.'[75]

8
Global Networks

There is no such thing as the mainstream, since the mainstream also relies on networks.

– Personal interview with Kailash Joshi

TiEcon, Silicon Valley, 2002

TiE began as a Silicon Valley-based ethnic organization. Its original intention was to connect Indian-American entrepreneurs who could use their common heritage as a platform for networking. A decade after its inception, however, it was clear that the association was undergoing a subtle yet profound transformation. TiE was becoming globalized.

The change was driven by practical necessity. For years the organization had expanded fairly haphazardly as a web of interconnected local chapters. Yet, as TiE grew in size and stature, explains Kanwal Rekhi, it needed to develop an institutional structure that could think globally, ensure a corporate identity and maintain the quality of the TiE brand.[1]

Due to this transformation, TiE began to shed its ethnic roots. Concerned that characterizing itself as an Indus organization might make people reluctant to join, TiE ceased to define itself solely as 'The Indus Entrepreneurs'. Today TiE stands instead for 'Talent, Ideas, Enterprise'. 'Our ethnic identity has served us very well so far', says Kailash Joshi, 'but we don't want it to be impeding our growth'.[2]

This current direction has its critics, as some members feel that going global will dilute the organization's original goal. There is little hesitation, however, among TiE's founders. Increasingly aware of the presence of more and more nonIndus people, and of more chapters opening in the nonIndus world, they have come to believe that their association is no longer typified by a South Asian ethnic identity. Instead, they prefer to see TiE as reflecting 'the culture and value system of the Silicon Valley'.[3] What holds TiE together, they explain, is its 'dedication to cultivating an entrepreneurial ecosystem, and anybody and everybody who wants to be a part of that is welcome'.[4] 'Our mission', states Mr Joshi, 'has no ethnicity.' Mr Rekhi concurs: 'We have always been open. We have never limited our attendance to Indians or Pakistanis. We have never declined admission to anyone.'[5]

When questioned about whether they were concerned that TiE might lose its ethnic character, both Mr Rekhi and Mr Joshi appeared unmoved. 'We are not going to lose it', said Joshi, 'it's there. And by the way even if it's lost we will have created a pretty phenomenal organization. So who cares?'[6] Kanwal Rekhi is equally adamant. 'I am most proud', he states, 'that TiE has pushed its own envelope of thinking and assumed a broader role for itself to foster socio-economic development globally'.[7]

This lack of concern, however, is in many ways unsurprising. Since, even when it defined itself in terms of its ethnic roots, TiE never really did fit the supposed ideal of an ethnic economy. Rather, it has thrived through its ability to synthesize the links and contacts of a supposedly marginal culturally based network with qualities of openness, transparency and meritocracy that are generally associated with the modern mainstream.

Becoming mainstream?

From its inception TiE has embraced the values of so-called Western modernity. Their brochure outlines a philosophical framework that is based in these 'mainstream' ideals. TiE's aim, they write, is to create an open, inclusive and transparent organization; to provide positive leadership and role models; to emphasize value creation through informed entrepreneurship; to maintain high ethical standards; to remain socially responsible; to display rigorous, intellectually honest behavior; and to pursue a modern, scientific, forward looking approach. With meritocracy as a ground rule they vow not to tolerate pettiness, divisiveness, insider privilege or corruption.[8]

This framework was generated through TiE's experience in the world of American IT. Adopting these values, however, is also the expression of a hope and ideal for India. One of TiE's goals, says Rekhi, was 'to define a new India which was modern and forward looking'.[9] This aspiration is not restricted to the Indian diaspora alone. It has also infected the country's own indigenous cyberculture.

IT companies like Infosys – together with both foreign and diasporic influence – have fundamentally changed India's traditional business culture. The custom in India was for businesses to be run by family-owned firms. These firms tended to rely on a network of social and business relations – what the Chinese call 'guanxi' – in order to operate efficiently. Moreover, in the traditional family-owned business, authority was centralized not only in the owner but also in the owner's family, which passed on leadership from generation to generation.

The IT industry, however, as Professor Anand Patwardhan points out, has produced a 'mold and pattern that is really very different'.[10] These new high-tech companies have emerged in large part from a new business class. Gurcharan Das, for example, points to a 1999 Business Standard list of 100 Indian billionaires, which states that 8 of the top 10 were new entrepreneurs[11]. These entrepreneurially owned, professionally managed IT businesses have provided a counter example to the connections and patriarchal authority of the old industrial families.

This transparent, entrepreneurial ethic has spread throughout the IT industry and now applies not only to the new companies but also to the older, more established corporations. Both Wipro and TCS, for example, put great emphasis on openness and integrity. Azim Premji, chairman of Wipro and one of the richest men in India, has even publicly abandoned the custom of family succession:

> Neither of my sons are part of the succession plan. That's not to say that they will never hold this position, but they aren't part of the plan at this time... At the end of the day, this is a large complex organization with a market cap of 8 billion. The worst thing you can do to the investor community, which includes me, is to get the wrong person in leadership. It's much cheaper to give my son 500 million and tell him to go and blow it than to get him into a leadership position where he destroys half the value of the company.[12]

It is Infosys, however, more than any other company, which embodies the new business ethic that IT has introduced to India. In a country renowned for its rampant corruption, Infosys has a reputation for corporate governance that would put many of its Western counterparts to shame. It is frequently heralded as 'India's most admired company' and is said to be 'fanatic about ethics', refusing to take bribes, paying all their taxes and sharing wealth.[13] Infosys, writes *Fortune Magazine* in an article praising Mr Murthy and Mr Nilekani as Asia's businessmen of the year, 'may be the only company in the world to publish financial statements in accordance with eight countries (Australia, Britain, Canada, France, Germany, India, Japan and US)'.[14] In an industry notorious for absurdly inflated salaries – especially in top management – 'chairman Murthy and CEO Nilekani are paid US$44,000 and US$42,000 annually respectively... [They] eschew even the most modest perks, reimbursing the company for personal phone calls and both are widely admired for leading normal middle class lives'.[15] Perhaps most crucially of all, reports *Fortune*, in a country where 'private corporations are often run by family fiefdoms, Murthy and Nilekani have kept their children off the payroll'.[16]

Beyond adopting these typically mainstream values, Indian IT professionals – both at home and abroad – have been very aware of the danger of sealing themselves off. In contrast to the typical cliquishness of ethnic economies, they are openly anxious about creating secluded communities that operate as isolated enclaves. Researcher AnnaLee Saxenian quotes one charter member of TiE speaking candidly about this inherent danger. 'This network just does not connect to the mainstream', he complains. 'If you look at the social gatherings that the TiE members go to, it's all Indians. There's nothing wrong with it...but I think if you don't integrate as much you don't leverage the benefit as much.'[17]

To counter this criticism, TiE seeks – as one of its defining features – to help members assimilate with the mainstream economy. Through contact, social networking and mentoring, TiE aims to teach its younger members 'how to think American' and 'provide the polished, professional product VC firms are searching for'.[18] In an interview in *The Entrepreneurial Connection*, Kanwal Rekhi comments proudly on TiE's success in this undertaking.

> The notion of Indian technology firms being featured on Nasdaq arose out of TiE. We are having a major impact and have been able to break several mindsets. For example, we have proven that Indians are just as good business people as they are technologists... A majority of business plans submitted to VCs in the Silicon Valley are now by Indians. What's more a significant number of projects funded today are headed by Indians. TiE can take the lion's share credit for this.[19]

The extent of this integration is further evidenced by the makeup of some – if not all – of TiE's local chapters. The Indian High Tech Council, the Washington chapter of TiE, for example, 'boasts a membership of 950 high level executives, the vast majority of whom are not Indian'.[20] Newspaper and magazine articles tell enthusiastically of the Council as one of the most successful networking groups in the region, with tendrils that clearly reach far beyond any particular ethnic group. *The Washington Post* describes one Council event in the following terms:

> Look! There's the British ambassador, entreating the fat wallets to invest in the UK. There's Mark Warner, would be governor of Virginia, who calls the council 'maybe the singularly most successful association that's taken place in the last decade.' There's Jamie Rubin, recent State Department spokesman and symbol of traditional Washington, smoking in the corner like an outcast teen. Occasionally you catch a glimpse of a turban or a purple sari.[21]

The discourse of ethnic economies, as AnnaLee Saxenian points out in her report *Silicon Valley's New Immigrant Experience*, tends to focus on immigrant populations that are concentrated in 'marginal' industries such as restaurants and small-scale retail. Despite their high level of self-employment these immigrants remain in the lower income brackets and, thus, continue to be of peripheral interest to the workings of the mainstream economy.

The Indian IT diaspora, on the other hand, which 'tends to be made up of the intellectual and commercial elite',[22] does not fit this mold. The recent wave of Indian immigrants to America is highly educated. Seventy-five percent of working Indians are college graduates and only 3 percent lack a high school education.[23]

As a group, they are also enormously prosperous. As Mira Kamdar points out, 'Indian immigrants in the United States . . . have been highly successful, earning the highest incomes per family of any immigrant group and enjoying, in general, a standard of living well above the American norm. The Indian-American community is one of the most prosperous in the country.'[24]

Even more crucially, the Indian IT diaspora works in the most advanced sector of the world economy. Their impact on the high-tech industries of Silicon Valley is so profound that *Fortune Magazine* does not exaggerate when it reports that 'without Indian immigrants the valley wouldn't be what it is today'.[25] Operating at the very core of the information age, the tight networks that produce the Indian IT ethnic economy – if that is what it is – can hardly deemed to be marginal to the modern mainstream.

The influence of networks

Notions of marginal capitalism sought to portray cultural networks as fundamentally distinct from the mainstream or general economy, which was characterized by impersonal relations and the strict laws that govern the bureaucratic machine. TiE challenges the rigidity of this distinction by showing that a supposedly marginal group is thoroughly infused with 'mainstream' traits. Yet, the opposite is also true. The mainstream is thoroughly permeated with the networks of dispersed cultural or ethnic groups.

All entrepreneurs use networks. Most start-ups are financed by family, and personal connections are just as applicable to modern capitalism as they are to its marginal undercurrents. In an article entitled *The Role of Networks in the Entrepreneurial Process*, Sue Birley writes 'that it's the informal contacts of friends, families and colleagues that are the main sources of help when it comes to finding equipment, space, and money but also advice, information and reassurance. In business', she concludes, 'the ability to build contacts and develop networks is fundamental'.[26]

These contacts are often based on ethnic ties, even though this is frequently overlooked by the tendency to assume that 'only nonwhites are ethnic'. In his article, *Thinking and Rethinking Ethnic Economies*, Antoine Pecoud stresses this point to critique the discourses of marginal capitalism. 'In a huge majority of cases', he writes, 'the term ethnic is used only to talk about immigrants and minority groups. The white majority is almost never called ethnic.'[27] Yet, the business partnerships forged on golf courses and Ivy League clubs are just as likely to be built upon cultural kinship as alliances that are sealed over curry or karaoke. Pecoud argues against 'guanxi' being employed to define 'the spirit of Chinese capitalism'. 'What exactly is Chinese here?', he asks. 'The use of personal relationships – whatever they are called – and of cultural affinities between businessman characterize nearly all economies, not only the Chinese.'[28]

The study of ethnic economies has produced valuable insights about how immigrant entrepreneurs function, and about the business webs of diasporic populations. Theoretically, however, the stress put on the distinction between the marginal and the mainstream has overshadowed the fact that there really is no such thing as a general economy. Since 'all groups are ethnic...all economies are ethnic as well'.[29] 'The notion of a unique and general economy with a single universal logic', writes Pecoud, 'is a fiction.'[30]

In the information age, networks are not only cultural but also geographic. This is especially true of Silicon Valley, a networked region, through which digital technology – which itself constructs a network – gets produced. The advantage of Silicon Valley's networked structure is illustrated by AnnaLee Saxenian in her study, *Regional Advantage*, which compares the Valley with one of America's older and more traditional zones of technological production – Boston's Route 128.[31]

Steeped in New England's formal and conservative business culture, Route 128 consists primarily of interiorized, insular firms that function with bureaucratic management and a strict hierarchical order. These firms are dominated by 'vertical integration and corporate centralization'. This creates, according to Saxenian, a 'culture of secrecy' in which highly guarded isolated companies – which insist on absolute loyalty – seek complete self-sufficiency and do not develop strong connections with the external world.[32]

Silicon Valley, on the other hand, is a fluid, highly flexible and technologically dynamic environment in which people are accustomed to taking risks, leaving established careers for start-ups, for example, or moving from one firm to another. 'Almost everyone in the Valley is from somewhere else', says Kanwal Rekhi, 'either in the US or abroad. Unlike the more established centers of technology on the East coast, the Valley has no traditions to limit its imagination.'[33]

Stanford, the main research university in the area, rejected the ivory tower approach of Harvard and MIT, and sought to integrate research with local businesses. At the same time, firms in the Valley substituted vertical integration for a decentralized system that spreads the cost of developing new technologies by linking up with outsiders. In addition, the Valley is filled with restaurants and coffee bars where workers – often from competing firms – come to socialize and talk about business and technology. Together with the large number of social and professional networks in the area, this 'informal communication and collaborative practices create a level of openness in which companies compete intensely while learning from each other'.[34] In this network-based culture, writes Saxenian, 'entrepreneurs came to see social relationships and even gossip as a crucial aspect to their business'.[35]

Even among the big companies that make their home in Silicon Valley, there is a movement away from centralized control toward open decentered management styles. It was in Silicon Valley, Saxenian points out, that companies like Hewlett Packard pioneered a new business culture, which emphasized flatness and the joint responsibility of the team. Strategies such as 'unplanned discussions' and 'management by wandering around' were used to create an open and participatory environment, which actively sought to break down the hierarchies between employers and employees, and 'worked to facilitate the free flow of information and ideas'.[36]

It is tempting, especially given the above comparison, to assume that the influence and pervasiveness of networks is new; a result, perhaps, of the peculiar traits of the information age. Capitalism, however, has, from its inception been based on the flows and connections of networks. At their origin these networks were produced from the webs created by dispersed ethnic groups.

The influence of these transnational ethnic webs constitutes, according to Joel Kotkin, 'one of the critical elements in the evolution of the global economy'. Where they appear, he writes, 'new combinations of technology, industry and culture flourish. When they leave, by choice or through compulsion, the commercial lifeblood, more often than not, runs dry.'[37]

Modern capitalism itself is a product of the British diaspora. It emerged as sea-faring merchants poured out of the UK. An 'essentially mercantile' population intent on 'business expansion', this diasporic culture spread throughout the world seeking not so much 'to export a civilization but rather to open networks of commerce and trade'.[38] In the process they established the standards and methods, rules and regulation through which the world economy now functions. The values of modern capitalism – 'including the idea of an individual business-man backed up by the force of law, and an independence from informal personal relations'[39] are a product of this diasporic flow. Still, 'today it is the diaspora of the British that constitutes the core of modern world society'.[40]

Thus, if the mainstream has a culture, it is – as proponents of the idea of Westernization believe – no doubt that of the English-speaking world, which dominates global trade and technology. The question is, is this culture best described by theorists of modern capitalism, who were haunted by visions of the industrial and bureaucratic machine? Or rather, is it best to evoke another machine when describing global culture; the network, with its diasporic populations and flows of communication, trade and exchange?

The Anglosphere

In the period following the Iraqi War of 2003, the concept of an 'Anglosphere' began to circulate in the journals and – especially – the blogs of the World Wide Web. The idea of the Anglosphere contends that English culture, that is to say the culture of the English language, has less to do with the Anglo Saxon people than with certain traits, values and ideals. In *An Anglosphere Primer*, probably the key Anglosphere text, James Bennett outlines the characteristics of this concept.

First, the Anglosphere is networked, rather than hierarchical. Its structure corresponds to what Bennett calls a 'Network Commonwealth', a cooperative unity built through alliances, trade blocks and joint security arrangements. The Anglosphere 'is polycentric and collaborative', writes Bennett, 'befitting an era in which the network, not some plan, is the ruling paradigm'.[41]

Due to this networked structure, the Anglosphere is more attuned to the wanderer – or, to use Benett's term, the sojourner – than the citizen. 'In the Machine Age', he writes, 'individuals were citizens of one nation-state and [they] resided, worked, and paid taxes within that state. The only way to change that status was to give up citizenship in one nation, move to a new nation and adopt residence, employment, and citizenship there'.[42] The 'Network Era' contrasts this Machine Age model of immigration with the mobility of the 'sojourner'.

The sojourner does not identify with the nation, and therefore 'does not seek to fill a citizen's slot'.[43] Rather, they move from country to country, wherever work and opportunity appear. Sojourners maintain their links virtually and are, thus, more intimately tied to cyberspace. 'As humans cease to be inhabitants and economic actors solely of physical space', writes Bennett, 'we begin to have an "amphibious" existence split between physical space and information space. Each space has its own rules and realities, and the sojourner is the person who helps tie the two together.'[44]

Finally, due to its network structure and tendency to favor personal transnational movement, the Anglosphere has little to do with its origins in the West. 'Anglospherism is assuredly not the racialist Anglo-Saxonism', insists Bennett. 'It is a memetic, rather than genetic, identity.'[45] To be part of the Anglosphere requires 'adherence to fundamental customs and values' like freedom, individualism and the rule of law rather than any racial unity or loyalty to a particular ethnic group.

> Those who come to use the language and concepts of the Anglosphere (and further their evolution) are the memetic heirs of Magna Carta, the Bills of Rights, and the Emancipation Proclamation, whatever their genetic heritage. 'Innocent until proven guilty' now belongs to Chang, Gonzales, and Singh, as well as Smith and Jones.[46]

The rise of the Anglosphere, therefore, does not imply the Westernization of the world, precisely because the culture of the English language no longer belongs to the West. 'The educated English-speaking populations of the Caribbean, Oceania, Africa and India', writes Bennett, 'constitute the Anglosphere's frontiers.'[47]

The world's link language

English's role in contemporary India is dependent on the language's ability to detach itself from its Western roots. English has survived in postcolonial India precisely because it has been deracinated and deterritorialized. As Braj Kachru writes, 'when English is adapted to another culture – to non-English contexts – it is decontextualized from its Englishness (or Americanness)'.[48]

These factors have allowed English to be adopted as India's intranational language. They have also contributed immensely to the success of English as an international tongue. For a language to be truly global it is essential that it be unhinged from any specific concrete culture. A 'lingua franca should not be associated with a particular ethnic group, religion or ideology', writes the political scientist Samuel Huntington. 'In the past English had many of these associations. More recently English has been de-ethnicized.'[49] No longer associated with any particular country or culture, English provides a neutral ground upon which intercultural communication can occur.

It is essential for any global language to satisfy this need for connections. According to Kachru, 'a universal language is one which, in its various forms and functions, is used by a large portion of the human population for easy communication between peoples of diverse cultural and language backgrounds'.[50] English, with its capacity to absorb, mutate and adapt, is able to fulfill this task both in India and throughout the world.

Since it operates as a link language, global English does not coincide with the spread of a homogenous universal civilization. Rather, it is precisely what enables cultural difference to persist. As Huntington argues, the need to adopt a single language for intercultural communication undermines the very idea of an advancing monoculture since it 'presupposes the existence of separate cultures . . . A lingua franca is a way of coping with linguistic and cultural differences', he writes, 'not a way of eliminating them.'[51] The English language 'which people from different language groups and cultures use to communicate with each other'[52] helps to maintain and, indeed, reinforces people's separate cultural identities. 'Precisely because people want to preserve their own culture', writes Huntington, 'they use English to communicate with peoples of other cultures.'[53]

Yet, the fact that English has been dissociated from any particular ethnic, local or regional culture does not mean that it has no culture at all. The plurality of English, its distinct and diverse uses in local contexts, and its ability to function as a medium for cross-cultural exchange is precisely what enables the language to construct a singular global cultural network built out of multiplicity, variation and change.

The other tongue

English has never been a culturally pure language. It arose among the Indo-Aryan steppe nomads and its roots are a hybrid mix of diverse cultural flows. Born amidst the exchange of a variety of people and tongues including Anglo-Saxon, Celtic, Norse, French, Latin and Greek, English has always grown from the bottom up.

Moreover, English is an open language that is not controlled from above. Unlike French, for example, it has no central agency or organized authority.[54] Eschewing all forms of cultural protectionism, English has no government body or other authority whose job it is to control language change and produce a standardized model. This open policy has been so successful that even today a single authoritative Standard English simply does not exist.[55]

English thrives then by spreading from below, creatively absorbing new words, which it borrows from each culture it encounters. Shakespeare, the language's most famous writer, is notorious for making up words and phrases. Today's English draws on everything from American street slang, to Jamaican dub, to Asian MTV. To quote from the book *The Story of English*:

> The great quality of English is its teeming vocabulary, 80 per cent of which is foreign-born. Precisely because its roots are so varied – Celtic, Germanic (German, Scandinavian and Dutch) and Romance (Latin, French and Spanish) – it has words in common with virtually every language in Europe: German, Yiddish, Dutch, Flemish, Danish, Swedish, French, Italian, Portugese and Spanish. In addition, almost any page of the Oxford English Dictionary or Websters Third will turn up borrowings from Hebrew and Arabic, Hindi-Urdu, Bengali, Malay, Chinese, the languages of Java, Australia, Tahiti, Polynesia, West Africa and even from one of the aboriginal languages of Brazil.[56]

The openness of English has not only created new words, but it has also spawned whole new dialects. 'The people who speak English throughout the world', writes Huntington, 'also increasingly speak different Englishes.'[57] These 'new Englishes' tend to go by hyphenated labels such as Jamaican-English, Hong Kong-English, and – when the connection grows more intimate – even more hybrid terms such as Chinglish (in China) and Hinglish (in India).

Native speakers have a tendency to dismiss, deride and mock these local varieties. Purists are alarmed at what they perceive as the language's decay, corruption and death, and generally dismiss any and all deviations as mistakes. Braj Kachru – whose work focuses on these new Englishes – argues that linguistic studies often reinforce these prejudicial tendencies, believing that 'there exists in the usage of a native speaker both a unity and a hierarchical superiority'.[58]

'Linguists, perhaps especially American linguists, have long given a special place to the "native speaker" as the only truly valid and reliable source of language data.'[59] Yet, English's non-native speakers are beginning to outnumber its native speakers, and 'much of the world's verbal communication takes place by means of languages which are not the user's "mother tongue", but their second, third or nth language'.[60] Though the relative linguistic homogeneity of China, with its vast population, makes Mandarin the most widely spoken language in the world, 'English is No 1 when those who speak it as a second, third or fourth language are counted.'[61]

In his attempt to shift focus away from the native user, Kachru emphasizes the creative mutations and linguistic innovations that proliferate among non-native users. His work repeatedly stresses the difference between a deviation – a new variety of English grown out of a particular context and culture – and a mistake. Because English lacks any central policing authority, it is the users themselves who decide which of these linguistic changes the English language can absorb. English speakers around the world are, thus, continuously shifting the language to fit their own particular culture, context and way of life.

The plethora and diversity of non-native speakers whose 'use varies from broken English to almost native (or ambilingual) competence',[62] has led Braj Kachru to call English 'the other tongue'. According to Kachru it is this 'otherness' of English, its deterritoriality and transnationalism, 'which has actually elevated it to the status of an international (or universal) language'.[63] 'By their geographical distribution, numerical strength and varied use of English, the non-native users have made English, as it were, a window on the world.'[64]

No place gives us better access to this window than India, where English has such an entrenched and complex position that it is no longer clear whether it is a native, non-native, foreign or indigenous tongue.[65] It is clear from looking at the example of English in India, that English is the language of a global culture, not because this culture is homogenous but because English, like the cosmopolitan culture to which it belongs, thrives on variation, diversity and change.

Indianization of English

The marriage of English and the languages of India has created, in Anthony Burgess' words, a 'whole language, complete with the colloquialisms of Calcutta and London, Shakespearian archaisms, bazaar whinings, references to the Hindu pantheon, the jargon of Indian litigation, and shrill Babu irritability all together. It's not pure English, but it's like the English of Shakespeare, Joyce and Kipling – gloriously impure.'[66]

In India in 1999 one of these impure phrases circulated around the country as a pervasive cultural meme. The phrase came from a slogan for a Pepsi ad. It read 'Yeh Dil Mange More'. As a foreigner in India who does not speak the local language, it took a while to understand that the slogan was a Hinglish phrase. 'Yeh Dil Mange More' translates as something like 'this heart asks for more'.

The ad campaign was glitzy and successful, but the slogan had a viral contagion of its own. Magazine columnists used it in lamenting the slow pace of economic reforms. The phrase fed into the hype-driven worlds of cricket and Bollywood, and was absorbed into the buzz surrounding IT. Most striking of all, however, was when it surfaced in association with the Indian military victory in Kargil. Newspapers reported that as soldiers pushed back the enemy and captured the key location of Tiger Hill, a young captain – later killed – stood on the conquered land and shouted out the Pepsi slogan: Yeh Dil Mange More.

Indian culture is synthetic. Built on layers and layers of alien influence and trade, it has always absorbed outsider influences and been strengthened by exchange. English – like Sanskrit and Persian before it – has had a deep influence on Indian life. Yet, the absorption of English in India is matched by the equally profound influence India has had on the English language. For over two hundred years, says the author Raja Rao:

> English has been used by Indians to serve typically Indian needs in distinct Indian contexts. As long as we are Indian – that is not nationalists, but truly Indians of the Indian psyche – we shall have the English language with us and among us, and not as a guest or friend, but as one of our own, of our caste, of our creed, our sect and of our tradition.[67]

The Indianization of English has resulted in two fairly distinct phenomena, which can be loosely differentiated through the terms Indian English and Hinglish.[68]

Indian English is a dialect, a particular form of English with its own particular pronunciation, rhythm and – to a certain extent – vocabulary.[69] 'It is the means through which the tempo of Indian life is infused into... English expression.'[70] This creative mutation of the English language began during the Raj. As early as 1886 *Hobson Jobson: A Glossary of Anglo-Indian Words* catalogued thousands of Indian words and phrases that had been adopted into English, including such common terms as curry, veranda, shawl, bamboo and monsoon.

Hinglish is more obviously hybrid, striking and inventive. It is used to describe the process in which a speaker switches back and forth between an Indian language and English. This process of punctuating sentences with English words and phrases is technically called code-mixing and is beginning to be studied as a language of its own. While Indian English is only spoken by a fairly small elite, Hinglish is much more widespread and can be found in 'publicity blurbs, class room interactions, public addresses, TV and Radio Interviews'.[71] The use of mixed code characterizes the verbal behavior of practically all educated Indians in all informal and semi-formal situations in different domains. This has had the consequence of making 'English much more pervasive and functionally relevant as far as day-to-day common usage is concerned'.[72]

India's English speakers have long recognized that the English language is not the sole property of England or North America but instead belongs to anyone who happens to use it. To quote from the Official Language Commission of 1956: 'It is not suggested that English be rejected merely because it is a foreign language for we entirely agree that a language is not a property of any particular nation, and obviously it belongs to all who speak it.'[73]

The fact that a Hinglish ad slogan can spread through a local culture embodying everything from the most fervent nationalism, to Bollywood hype, to IT success, illustrates the extent of India's participation in the language of global culture. The slogan itself suggests that, as Kachru writes, India has 'played the age old trick on English too, of nativizing and acculturating it – in other words of Indianizing it'.[74] To quote Salman Rushdie, India's most famous English writer 'English, no longer an English language, now grows from many roots; and those whom it once colonized are carving out large territories within the language for themselves. The Empire is striking back.'[75]

Technological mutations

Those who seek to challenge the idea of a single global culture maintain that the technologies of globalization are already undermining the dominance of English. They point to such technological developments as automatic translators, foreign language software and the introduction of Unicode, a worldwide character set that is quickly becoming a universal standard. Unicode – unlike previous encoding schemes like ASCII – can represent tens of thousands of characters enabling programs and text files to be written and read in the majority of the world's diverse tongues. This linguistic pluralism is growing even among the most English – or American – of technologies. Barbara Walraff, for example, calls attention to the fact that Windows Millennium is available in 28 languages and though, what is currently on the Internet may be 80 percent in English, the fastest growing group of users are non-English language speakers. Yet, it is because of its ability to adapt to the technological networks that Global English continues to grow – and spread.

When mobile phones first entered the market place no one expected that the alphabet encoded on the numeric keyboard would be of much use. Tiny keys and a screen not much bigger than your thumb made for a clumsy and difficult interface. Initially little was done to promote text-based applications. In fact the industry was so blind to the possibilities of using the phone as a writing instrument that – for a while at least – text messages were free.

Yet, despite, or even because of these apparent shortcomings, SMS (short messaging services) has grown exponentially. It is now the preferred mode of communication for millions of people worldwide who – using their phones as keyboards – now send billions of messages every month.

Relatively unknown in North America,[76] this phenomenon is already commonplace in Europe and Asia. Yet, though it is being spearheaded by the non-English speaking world,[77] many of the world's SMS messages are sent in a quasi-English jargon, the strength of which is causing a radical mutation in the English language itself.

The rigid constraints of SMS technology – the difficulty of typing, combined with the fact that most phones restrict messages to no more than 160 alphanumeric symbols – has created a whole new dialect of English. Though it is still too early to fully catalogue, its traits include: the use of single letters or numbers for words ('C' for 'see', 'U' for 'you', '4' for 'for'), inventive alphanumeric abbreviations ('gr8' for 'great', 'l8r' for 'later') and terms composed solely of non-linguistic signs ('*$' for

'Starbucks'). A common text message thus reads: 'C U l8r @ *$' (see you later at Starbucks).

Though linguist purists may be horrified, it is because English is at the forefront of such techno-linguistic innovations that it remains the world's global tongue.

In October 2002 the English language magazine, *India Today*, ran a cover story entitled *Love, Sex and SMS*. The article used illustrations that parodied images of the Kama-Sutra. The cover showed an Indian couple in traditional dress seated back to back on a Persian rug. The couple gazed out lovingly into the distance. In each of their hands was a cell phone.

The article began with a statistic only comprehensible to those familiar with Indian-English. 'Over 2.5 crore SMS are sent by four lakh cell phone owners daily, an average of 60 messages a day.'[78] The article went on to tell how SMS technology, which is used by more woman than men, is shifting the communication patterns between friends and lovers. It thus detailed the profound and intimate shift text-based digital technology is having on Indian life.

By far the majority of these messages are sent in English, Hinglish or some strange mutation of both. The fact that new varieties of English are continuing to flourish in the most innovative sectors of one of the world's most populous nations makes it clear that English – in all its varieties, will retain, at least for the foreseeable future – its unrivalled status as the language of globalization.

Yet, the fact that everyday English is peppered with Indian words and phrases (pepper itself is ultimately derived from the Sanskrit *pipalli*) shows that this culture is far from homogenous, but is built instead on interlinking webs of communication, and flows that construct the worldwide networks of exchange.

9
Zero Logo

> The assumption that modern society must approximate to a single type, the Western type, that modern civilization is Western civilization and that Western civilization is modern civilization...is a totally false identification.
>
> – Samuel Huntington, *The Clash of Civilizations and Remaking of World Order*

> So what is the mystery behind India's success in providing efficient software solutions. Some say that it is the mathematical ability of Indians, others quote 'after all it was India that invented the numeral zero'.
>
> – Dewang Mehta, *The Software Industry in India: A Strategic Review*

Branding India Inc.

India's IT industry had no greater publicist than the late Dewang Mehta, former president of Nasscom. Mehta contributed immensely to the hype surrounding digital technology in India. He appeared frequently in the national press alongside such celebrities as Bill Gates and Bill Clinton, and he tirelessly promoted the country as an emerging 'software super-power'. With his Elvis Presley haircut and his near fanatical evangelizing, Dewang Mehta had, by the time of his sudden death in 2001, reached superstar status in the world of Indian IT.

Kanwal Rekhi, in his obituary to his friend, praised Mehta as a marketing genius, and there is no doubt that under his leadership, Nasscom saw branding India Inc. as one of its primary tasks. Nasscom was formed in 1988 as a 'catalyst for the growth of the software led IT industry in India'. Initially its focus was on building the 'India brand in software' in order to attract foreign business and promote the country's IT industry abroad.[1] By the turn of the millennium this goal had basically been achieved. 'If you go to the US and UK', says Sunil Mehta current vice president of Nasscom, 'every taxi driver more or less has heard of Indian IT.'[2]

Though the original emphasis on brand recognition has shifted, branding India remains a Nasscom priority. This was very much in evidence at their 2003 annual industry-wide conference. The stylish event was held at the Hotel Oberoi located in Mumbai's Nariman point, overlooking the Arabian Sea. Upon registration, delegates were given, along with their stacks of glossy brochures, a mousepad decorated with a line drawing depicting the silhouette of an archer. A postcard-sized handout explained that the silhouette was the logo for Nasscom's latest marketing campaign. The drawing was of the great prince Arjuna, a famous mythological character from the ancient Indian epic, Mahabharat. Underneath the drawing was the tagline 'India is IT'.

Dewang Mehta was replaced by Kiran Karnik who came to his job as president of Nasscom after running Discovery Channel on Indian TV. His continued commitment to branding is revealed in an article – posted in the 'media' section of Nasscom's website – entitled *India and IT: like France and Wine*. 'I want to promote the India Inc. brand abroad,' the article quotes Karnik as saying. 'I want to make India and IT as synonymous as France and wine or Switzerland and watches.'[3] The article continues:

> With other Asian countries trying to emulate India's software success, Karnik says the best way for India to stay ahead is to compete not on cost but on quality, service, and productivity, things he associates with a brand. As a software executive in India, he says, 'you want Indian IT to stand for something so that you don't have to worry about China or the Philippines undercutting you on cost. The only way we can compete is to move away from the dimension of cost to other areas where we have the advantage.' Thus, Indian companies need to focus on areas such as 'assurance of consistent quality, adding bells and whistles, [building] the relationships, and the comfort factor,' says Karnik. 'Build in the attributes of a brand. In the long run, this will be crucial to us'.[4]

Nasscom's branding strategy, which began with Mehta, is to link the marketing of India's software industry to the promotion of the country and its culture. This tactic is not at all uncommon. It often happens that the promotion of a particular product or industry becomes intertwined with the project of branding a nation. When Rituraj Nath, in a 1999 interview, described Nasscom's aims he did so by comparing Indian software with German cars. The perceived precision and efficiency of German culture, he explained, helps advertise the German automobile industry. This same process is at work in the case of French wine and Swiss watches. Everybody knows that French culture places great emphasis on its fine taste in food and drink. In the same way, the Swiss have a reputation for being meticulously well ordered, so it is no surprise that the country is famous for its banks and watches.

As these examples make clear, the practice of national branding is intimately connected with a nation's culture. Branding a nation, however, does not consist simply of distilling convenient, preexisting cultural traits. On the contrary, if done well, the project of national branding can be intensely creative, contributing to the production of culture itself – not by making up what does not exist – but rather through a process of reinvention and rediscovery that forges a new relationship to past traditions and cultural roots. A recent example is the 'Cool Britannia' and 'Think Britain' campaign, which sought – perhaps not altogether successfully[5] – to shift England's image abroad by promoting the eclectic, creative, racially mixed contemporary reality of Britain as opposed to its more stale and traditional associations of the Queen, scones and tea.

In so far as it is engaged with transforming a nation's image, branding often exists at culture's cutting edge. This is because the practice of branding – at least when most creative – views culture not as a set of static customs that must be faithfully reproduced, but rather as something to be constantly shaped through a dynamic process of invention and change.[6] In India's case, the attempt to construct itself as an IT superpower involves a positive renewal of its own indigenous culture, a conscious rebranding of its population and popular myths. Nasscom – and the world of Indian cyberculture to which it belongs – are engaged in a process of cultural creation, which recognizes that the past is not just sitting there waiting. It has to be reconstructed.

The project of branding India works by reconceptualizing those things which are stereotypically Indian in such a way as to show that, no matter how ancient, they were always closely intermeshed with the digital technology of today. Gurus on mountain tops and yogis in caves are no longer viewed as an escape from the practical realities of everyday life – as theorists like Weber had argued. Rather, the world of meditation and distant spiritual contemplation, with its supposed lack of materialism, is now rethought so as to be intimately connected with the future of the information age. Ascetic mystics, sadhus and snake charmers with few practical skills but an access to the 'otherworldly' have been replaced by abstract thinkers and talented mathematicians; in short the ideal 'knowledge workers'.

'India's forte is in the knowledge based industry', said Kailash Joshi in an off-hand remark made while sitting in a conference room at TiE headquarters in Silicon Valley. 'Chinese and Japanese they have manual dexterity. Indians are not blessed with that. My strong conjecture is that the tradition of India is not a materialistic culture. Manufacturing has a materialistic angle to it. We have a knowledge based tradition that goes back 6,000 years.'[7]

The marketers of India Inc. seek to 'revive' this tradition by reclaiming India's mathematical heritage, in order to promote the natural affinity that exists between Indian culture – particularly its numerical culture – and the codes and patterns of IT. 'What is the mystery behind the success of the Indian software industry?' wrote Mehta, echoing this Chapter's epigraph, 'Some praise the Indian intellect, which invented the numeral zero, pointing out that it was India that gave the world the decimal system.'[8] From the knowledge of the Vedas, to the numerical patterns of Kathak dance, to the informational layout of Sanskrit mantras and meditational diagrams, Indian culture exhibits an overwhelming desire to play with numbers and a resultant numerical complexity and sophistication that is simply unmatched in the West. Indians themselves are proud of this seemingly inherent mathematical genius. 'The Indian mind', writes Gurcharan Das, 'reads math equations like poetry.'[9]

Nasscom's goal is to capitalize, quite literally, on this heritage. 'Historians would tell you', said Dewang Mehta in an interview with the *Financial Times* 'that it was India that invented the number zero. And if you look at the binary system, it is just zero and one. So if we'd kept copyright of the zero system, 50 per cent of the innovations in computers would have been attributed to India'.[10] Mehta's tone has a lighthearted humor, but the statement has a calculated impact. With the invention of zero, the Indian IT industry has found the perfect cultural legacy, an element of the ancient past that plugs directly into today's most advanced technology, and is, thus, actively engaged with a future yet to come.

The flows of global culture

Since the success of Naomi Klein's book, *No Logo*, branding has been of central concern to the anti-globalization movement. Protesters have traveled across the Americas, through Europe, to Asia and beyond, attacking global brands as the overt signs of the Westernization, or more precisely, the Americanization of the world. The Nike swish and the McDonalds golden arches are condemned as the crass markings of a corporate culture that acts as a Juggernaut,[11] crushing all indigenous production and local diversity that happen to get in its way.

Though the assaults of the protesters are often aimed at surface symbols, their argument, they claim, is with a much deeper and more fundamental global structure. Their criticism harks back to the theories of the 1960s. Branding is accused of contributing to a new international division of labor, which maintains power relations between core and periphery zones. While global companies hire highly paid marketers and consultants to do the creative task of building brands in the West, the actual production of these goods is outsourced to the periphery, where labor is cheaper and the economy is dominated by unbranded, low value goods.[12]

The branding of India's software industry, however, tells a different tale. Though India Inc. is a global brand, it does not emanate from the West, but is instead a positive assertion of – and by – the periphery. Culturally, 'India as a software superpower' challenges a monolithic notion of globalization, introducing in its place the idea of global culture as a network that flows in many directions at once.

In the outstanding study, *Many Globalizations: Cultural Diversity in the Contemporary World*, Tulasi Srivinas takes issue with the conclusions of Peter Berger, the project's editor, that 'cultural globalization is the movement of goods and ideas (cultural freight) from the West to the rest of the world'. In an essay entitled *A Tryst with Destiny: The Indian Case of Cultural Globalization*, Srivinas seeks to examine cultural globalization as a two-way process, focusing in particular on the 'cultural models [that] are increasingly emitted *from* India... Indian cultural artifacts are consumed all over the world: silk sari bedding is advertised at Bloomingdale's, Indian Jewelry and dress, henna tattoos, Darjeeling tea, and toe rings are bought every day by Europeans and Americans'.[13] From curry to Karma, elements of Indian culture – think New Age gurus, Ayurvedic medicine, yoga and meditation – have flooded into the West, influencing everything from dietary habits to philosophical speculation.

By far the most profound contribution that India has made to global culture, however, is in giving the world the decimal number system. There is no doubt that the invention or discovery of zero is one of the most important innovations in the history of human culture. It is, according to mathematical historian John McLeish, 'no less important than such feats as the mastery of fire, the development of agriculture, or the invention of the wheel, writing or the steam engine, [and will] always stand out as one of the greatest single achievements of the human race'.[14]

Georges Ifrah, in his tome *The Universal History Numbers*, argues that Indian numeracy has reached a state of 'mathematical perfection'. 'No further improvement of numerical notation is necessary, or even possible,' he writes. 'Once this discovery has been made, the only possible changes remaining could only affect – the choice of base . . . [or] the graphical form of the figures.'[15] 'It is impossible to exaggerate the significance of the Indian discovery of zero'[16], insists Ifrah, which 'gave the human mind an extraordinarily powerful potential. No other human creation has exercised such an influence on the development of mankind's intelligence'.[17]

Though cultural critics rarely speak about numeracy, there is, in fact, nothing as profound. Your number system is your culture, as Georges Ifrah writes 'to know how a people counts is to know what kind of people it is'.[18] In capitalist culture the decimal system determines both time and money and is, thus, fundamental in shaping day-to-day existence. Moreover, the Indian method of counting is one of the transcendental presuppositions of cyberspace. The development of computers would 'never have occurred without the place value system devised by the Indians . . . if the positional number-system with a zero had not existed', writes Ifrah, 'the problem of mechanizing the process of calculation would never have found a solution; still less would it have been conceivable to automate the process'.[19]

The notion that contemporary globalization can be equated with Westernization presumes that the world's myriad flows of trade and communication can all be subsumed under one overarching unity. Ultimately, at the most fundamental level, this is a Western idea. The privileging of unity that this notion involves rests on a culture – and numerical system – that prioritizes 'The One'. Paradoxically, the universality of this presupposition is highly questionable. Indeed, as we will see in the pages that follow, the trade and technology of globalization does not rest on the West's indigenous numerical culture. Rather, it is constructed out of a number system that came from the East, and begins with zero, not one.

The story of zero, from its origins in India to the resistance it met in the West, has its own tale to tell about globalization and the culture to which it belongs. This is the story of a periphery invading the core to create a global culture that operates as a flat network, which – like the numeral zero itself – has nothing at the center and is edges all around.

The origin of zero

Most historical accounts trace zero back to Babylonia of the 3rd century BCE, where it operated as a place value indicator, although not yet as a number.[20] As with most ancient cultures, Babylonians used a calculating machine – or Abacus. The notation of these calculations involved a rudimentary positional or place value system, which differentiated numbers according to the columns they occupied. The problem with this system is that it required some way of indicating an empty column. By around 300 BCE the Babylonians solved this difficulty by using what many believe are the first signs of zero when they began to draw 'two slanted wedges to represent this empty space'.[21]

The Babylonian zero had no fixed place on the number line, and therefore did not have a numerical value of its own. Nothing more than an empty space in the abacus, it was not yet 'understood as a number synonymous with "empty" and never corresponded to the meaning of "null quantity"'.[22] Without a place on the number line zero was only a digit, with no value of its own.

Though it had a variety of precursors,[23] most historians agree that the invention or discovery of our number system occurred in Ancient India, where the zero-to-nine decimal numerals first appeared. According to both Georges Ifrah and John McLeish, the richness and innovation of India's numerical culture depended on three great ideas. The first was the use of abstract number symbols. With these symbols each basic figure was given a 'graphical sign removed from all intuitive associations, which did not visually evoke the units they represented'.[24] Second was the use of a place value system 'in which the value of a number depends on its position in the units, tens, hundreds, thousands, and so on'. Finally, but, according to McLeish 'most important of all – and a milestone as vital in the history of civilization as the invention of the wheel',[25] was the use of a symbol for zero. What makes the Indian zero truly unique is that, for the first time, it shifted from being simply a placeholder – the mark of an empty space – to functioning as a true number. The Indian zero – unlike its predecessors – operated as both a notational marker and a numerical concept, as empty place and null value.

The sparks of these discoveries go as far back as Mohenjo Daro (2550–1550 BCE), where a simple decimal system seems to have existed. The fully operational zero, however, dates from much later. In his book on zero, John Barrow writes 'The earliest example of the use of the Indian zero is in AD 458, when it appeared in a surviving Jain work on cosmology, but indirect evidence indicates that it must have been in use as early as 200 BC.'[26] Four hundred years later, zero seems to have been fairly well established. A 6th-century poem, Vasavadatta, speaks of how the 'stars shone forth ... like zero dots ... scattered in the sky'.[27]

The concept of zero was perfected by the mid-7th century, when Brahmagupta (*c.* 628 CE), the famous Indian astronomer, detailed methods for adding, subtracting, multiplying and dividing by zero.[28] In addition to classifying the uses of zero as number Bhramagupta also developed the notion of negative numbers, whose existence depends on the fixity of zero on the number line.

Bhramagupta's writings showed, for the first time, the extent to which zero behaves in highly mysterious ways. This is already evident in the simple calculations of addition, subtraction and multiplication. With addition and subtraction, zero acts as nothing and has no effect. (Add or subtract any number from zero and it is always itself.) With multiplication, on the other hand, this lack of impact reverses itself as zero asserts its force by 'collapsing all numbers into itself'. Any number multiplied by zero is zero.

It is in division, however, that the force of zero's strange power is most clearly revealed. Division exposes zero's möbius strip like nature, in which its two components, nothingness and infinity, become interchangeable. Any number divided by zero is infinity – infinity divided by any number is zero. 'Zero is powerful,' writes Charles Seife, 'because it is infinity's twin.'[29]

Greek geometry versus Indian arithmetic

Though the origins of our number system are obscure, what is clear is that its invention – or discovery – depended on a particular numerical and philosophical tendency that was able to flourish in ancient India. 'It is remarkable', writes Florian Cajori in his book, *History of Mathematics*, 'to what extent Indian mathematicians enter into the science of our time. Both the form and the spirit of the arithmetic and algebra of modern times are essentially Indian and not Grecian.'[30] By branding India as the inventor of zero, the Indian IT industry has zoomed in on a most crucial – and decidedly non-Western – contribution to the networks of global culture.

Much of the literature in the history of mathematics has a Western bias that tends to assume all intellectual innovations have their roots in Ancient Greece. The puzzle as to why, in the two great classical civilizations of the West – Greece and Rome – there was no evidence of a zero sign is, therefore, given a great deal of attention. 'The great mystery of zero', quotes Barrow in one of the more blunt statements of this type, 'is that it escaped even the Greeks'.[31]

The most direct reason for this oversight was that in Ancient Greece number was tied to measurement. More precisely, while the crass materialism of commonplace calculation was left to the slaves, the philosophers and statesmen pondered the dimensions of shapes. The Platonic reverence for this particular mathematical science is made evident in the dialogue, *Meno*, when Socrates uses the measurements of a square to prove the ideal nature of the Forms to his slave. Even more stark is the fact that inscribed on the door to Plato's academy were the words 'no one shall enter here who does not know geometry'. The Greeks had a particular fondness for square and triangular numbers since they saw little difference between numbers and the shapes they represented. It is this 'equivalence of numbers and shapes that made the ancient Greeks the masters of geometry'. Yet, this equivalence also 'had a serious drawback. It precluded anyone from treating zero as a number. What shape, after all, could zero be?'[32]

Philosophers Deleuze and Guattari argue that 'Greek geometrism' is intrinsically linked to the transcendent and idealist philosophy that arose in the Greek city state. By tying mathematics to measurement, the Greeks operated with a numerical culture in which number 'served to gain mastery over nature, to control its variation and its movements, in other words to submit them to the spatio-temporal framework of the State'.[33] Proclus, in his *Commentary to the First Book of Euclid's Elements*, makes plain these connections between geometry, measurement, territory and the State. 'Geometry', he writes, 'was first discovered among the Egyptians. It originated in remeasuring their lands. This was necessary for them because the Nile overflows and obliterates the boundary lines between properties.'[34]

Deleuze and Guattari contrast Greek geometrism with an Indo-Arab 'arithmetism'. This arithmetic culture belongs to the nomads rather than the State. Arithmetism is deterritorialized. It frees number from measurement. Arithmetism, write Deleuze and Guattari, operates with a 'numbering number' that belongs to a 'smooth space' 'independent from metrics'.[35] They call this smooth space *nomos* and argue that it is exterior to Greek or Western logos. 'When Greek geometrism is opposed to Indo-Arab arithmetism', they write, 'it becomes clear that the latter implies a nomos opposable to the logos: not that the nomads do arithmetic or algebra, but because arithmetic and algebra arise in a strongly nomad influenced world.'[36]

This distinction between logos and nomos, the State and the nomads, runs throughout *Capitalism and Schizophrenia*, Deleuze and Guattari's most important work. We need not engage with the intricacies of their philosophy, however, to recognize the distinction they are trying to make. 'Very striking', writes Cajori 'was the difference in the bent of mind of the Hindoo and Greek; for while the Greek mind was pre-eminently *geometrical*, the Indian was first of all *arithmetical*. The Hindoo dealt with number, the Greek with form.'[37]

In India a numerical culture arose that was not based on geometry and logic – which tie number to measurement and form – but was instead arithmetical. Unlike the Greeks, the 'Indians did not see squares in square numbers or the areas of rectangles when they multiplied two different values. Instead they saw the interplay of numerals.'[38] Stripped of their geometric significance, numbers were able to operate as abstract signs with their own intrinsic properties. No longer used to merely measure objects, 'numbers had finally become distinct from geometry'.[39] Indian culture, which is passionate about numbers, developed a series of practices, from calculation to complex rhythmic art, that are based on the pure play of numbers.

Revering nothingness

The abstraction of zero was an anathema to the Western mathematical tradition that tied number to the measurement of space. It was not only mathematically, however, but philosophically and religiously too, that zero belonged to the East. In India, 'the creation of a numeral to denote no quantity or an empty space in an accounting ledger was', as Barrow writes, 'a step that could be taken without the need for realignment of parts of any larger philosophy of the world'.[40]

The Sanskrit word for zero is *shunya*, a word that existed long before the place value system. Shunya literally means void or empty,[41] yet these English translations miss much of the complexity of the term. In the *Dictionary for Indian Numerical Symbols*, which forms the appendix to Georges Ifrah's *Universal History of Numbers*, the entry for zero reads as follows:

1. The void (*Shûnya*). 2. Absence (*Shûnya*). 3. Nothingness (*Shûnya*). 4. Nothing (*Shûnya*). 5. The insignificant (*Shûnya*). 6. The negligible quantity (*Shûnya*). 7. Nullity (*Shûnya*). 8. The 'dot' (*Bindu, Vindu*). 9. The 'hole' (*Randhra*). 10. Ether or 'element which permeates everything' (*Âkasha*). 11. The atmosphere (*Abrha, Ambara, Antariksha*,

Nabha, Nabhas). 12. Sky (*Nabha, Nabhas, Vyant, Vyoman, Vishnupuda*).
13. Space (*Âkasha, Antariksha, Kha, Vyant, Vyoman*). 14. The firmament
(*Gagana*). 15. The canopy of heaven (*Gagana*). 16. The immensity of
space (*Innuendo*). 17. The 'voyage on water' (*Jaladharapatha*). 18. The
'foot of Vishnu' (*Vishnupada*). 19. The zenith (*Vishnupada*). 20. The
full, the fullness (*Pûrna*). 21. The state of that which is entire,
complete or finished (*Pûrna*). 22. Totality (*Pûrna*). 23. Integrity (*Pûrna*).
24. Completion (*Pûrna*). 25. The serpent of eternity (*Innuendo*).
26. The infinite (*Innuendo, Vishnupada*).[42]

Zero is a powerful concept as well as a number. In India, shunya – or
the void – was divinized in both these aspects as the ultimate cosmic-
religious aspiration. The Rig Veda teaches that 'in the earliest age of the
gods, existence was born from non-existence'. Indian culture 'regarded
Nothing as a state from which one might have come and to which one
might return – indeed these transitions might occur many times, without
beginning and without end'.[43] 'Nothingness', writes Seife, 'was what the
world came from, and to achieve nothingness again becomes the ultimate
goal of mankind.'[44]

Far surpassing 'the heterogenous notions of vacuity, nihilism, nothingness, insignificance, absence and non-being of Greek and Latin philoso-
phies',[45] the Indian concept of zero was more nuanced, rich, multiple
and dense. It has been possible, writes Ifrah,

> to distinguish twenty five types of *shunya*, expressing thus difference
> nuances, among which figure the void of nonexistence, of nonbeing,
> of the unformed, of the unborn, of the nonproduct, of the uncreated
> or the nonpresent; the void of the nonsubstance, of the unthought, of
> immateriality or insubstantiality; the void of nonvalue, of the absent,
> of the insignificant, of little value, of no value, of nothing, etc...[46]

Amongst the signs for zero was the dot or 'bindu'. The bindu is often
used as a structuring feature of Tantric and Buddhist diagrams. It is, as
Ifrah writes, the sign of 'the universe in its non manifest form and
consequently constitutes a representation of the universe before its
transformation into the world of appearances'.[47] The sacredness of the
symbol, and its use in meditational exercises show 'how the state of
non-being was something to be actively sought by Buddhists and
Hindus'.[48] Literally bindu is just a 'point', but in these sacred 'yantras
and mandalas this point symbolizes the potential or virtual energy of
the uncreated universe. 'By its motion', Barrow writes, 'a single dot can

generate lines, by whose motion can be generated planes, by whose motion can be generated all of three dimensional space around us. The bindu was the Nothing from which everything could flow.'[49]

Indian culture embraced zero not only as nothing, but also in its aspect as infinity. 'Where there is the infinite there is joy', reads the Chandogya Upanishad, 'there is no joy in the finite.' Amongst the Sanskrit words used to express zero is *Ananda*, literally 'Infinity'. In Indian mythology Ananda is the name of the serpent on which Lord Vishnu rests between creations. 'From a mythological, cosmological and metaphysical point of view, the zero and infinity have come to be united, for the Indians, in both time and space.'[50]

Unity and the West

What was accepted and revered in India was deemed alien and adamantly resisted – even banished – in the West. 'The biggest questions in science and religion are about nothingness and eternity, the void and the infinite', writes Seife. 'The clashes over zero were the battles that shook the foundations of philosophy, of science, of mathematics, and of religion. Underneath every revolution lay a zero – and an infinity. Zero was at the heart of the battle between East and West.'[51]

Western culture and civilization are generally understood as a synthesis between the Judeo-Christian biblical tradition and the rational philosophy of the Greeks. Both these strands, in their separate ways, forcefully rejected zero and all it implied. 'In Europe', writes Seife, 'zero was an outcast.'[52] 'Zero conflicted with the fundamental philosophical beliefs of the West, for contained within zero are two ideas that were poisonous to Western doctrine... These dangerous ideas are the void and the infinite.'[53]

The Greeks had too much respect for logic to think that nothing could be something. 'The great Parmenides', writes Plato in his dialogue *Theaetetus*, 'constantly repeated in both prose and verse: Never let this thought prevail, that not being is, but keep your mind from this way of investigation'.[54] The Greeks explicitly rejected zero – recoiling from the horror of the void – in favor of the supremacy of the One. Western philosophy from Zeno, to Parmenides, to Aristotle centered on this reverence for the unity of One. 'The whole Greek universe rested upon this pillar: there is no void. The Greek universe, created by Pythagoras, Aristotle, and Ptolemy, survived long after the collapse of Greek civilization. In that universe there is no such thing as nothing. There is no zero.'[55]

Similarly, 'in the Judeo-Christian tradition nothing was considered the antithesis of God. God's defining act was to create out of nothing. Nothing, was therefore a state without God – the chaos and anarchy that lay outside God's domain.'[56] Governed by the principle of the unity of God, this culture fed on the primal fear of void and chaos, a realm of dread before and beyond the order that God had imposed. To quote from *Paradise Lost*:

> Before their eyes in sudden view appear
> The secrets of the hoary deep, a dark
> Illimitable Ocean without bound,
> Without dimension, where length, breadth, and height,
> And time and place are lost; where eldest Night
> And Chaos, Ancestors of Nature, hold
> Eternal anarchy, amidst the noise
> Of endless wars, and by confusion stand.[57]

'The Hebrew tradition regarded the void as the state from which the world was created by the movement and the word of God', writes Barrow. 'It possessed a host of undesirable connotations. It was a state from which to recoil. It spoke of poverty and lack of fruitfulness; it meant separation from God and the removal of His favor. It was anathema...'[58]

The Christians went so far as to link zero with Satan. 'It is true', wrote the philosopher Leibniz, 'that the empty voids and the dismal wilderness belong to zero, so the spirit of God and His light belong to the all powerful-One.'[59] For the Church, the numeracy of India 'reeked of magic and the diabolical'.[60] 'Aquinas spirited the Aristotelian abhorrence of Nothing by viewing the creation of the world as an annihilation of Nothing in an act of Divine creative transformation.'[61] Augustine went even further, equated Nothing directly with the Devil: 'It represented complete separation from God, loss and deprivation from all that was a part of God, an ultimate state of sin, the very antithesis of a state of grace and the presence of God. Nothing represented the greatest evil.'[62]

Before Europe could accept the numeracy of the East, a profound cultural transformation had to occur. The West was forced to acknowledge the existence of the void and of the infinite. It had to embrace – as its underlying foundations – a numerical system that does not start with unity. 'Zero', writes Seife, 'clashes with the fundamental philosophy of the West.'[63] For the West to accept zero required that its conceptual universe be destroyed.

The spread of the Hindu-Arab numerals

Underlying Western religio-philosophical objections to the zero sign was a quite pragmatic concern. Indian arithmetic was much simpler than the alphanumeric system of Greece and Rome. This more 'democratic' form of arithmetic was a direct challenge to the primacy, and authority of the Church. Before the intrusion of this less complicated numerical culture, 'science and philosophy were under ecclesiastical control, they were obliged to remain in accordance with religious dogma and to support, not to contradict, theological teachings'.[64] It was much easier for the church to retain its power and privilege when the art of mathematics was available to only a few.

The Church had to struggle, however, to hold on to its monopoly on arithmetic. From the end of the 8th century, zero entered the world of Islam, and rapidly spread throughout the vast expanse of Arabic-Islamic civilization. 'Islam and Oriental Judaism were able to adopt zero under the guise of the primal chaos before and beyond creation.'[65] As early as the 9th century CE, the Indian zero found its way to the threshold of Europe through Spain via this channel of Arab culture.[66]

It was not for another 500 years, however, that zero and the numerical culture it supported began to seep through Europe's well-protected walls. For Europe, the 'dawn of the modern age did not really occur until Richard Lionheart reached the walls of Jerusalem. From 1095 to 1270, Christian knights and princes tried to impose their religion and traditions on the Infidels of the Middle East.'[67] The spread of the Hindu-Arab numerals – as they now came to be called – was greatly assisted in 1202 when Fibonacci published Liber Abaci. 'From 1202', writes Ifrah, 'the trend began to swing in favor of the algorists, and we can thus mark the year as the beginning of the democratization of number in Europe.'[68]

Resistance to the new methods, however, was not easily overcome. Natural inertia and conservatism 'held up the wholesale introduction of the Indo-Arab system among the majority of merchants until well into the sixteenth century'.[69] The West not only refused to teach the new system, they also passed laws attempting to abolish it. 'In 1299 a law was passed in Florence forbidding the use of zero. The reason was fear of fraud.'[70] Ifrah describes the abysmal state of arithmetic in 15th and 16th century Europe by use of the following anecdote:

A wealthy German merchant, seeking to provide his son with a good business education, consulted a learned man as to which European institution offered the best training. 'If you only want him to be able to cope with addition and subtraction,' the expert replied, 'then any French or German university will do. But if you are intent on your son going on to multiplication and division – assuming that he has sufficient gifts – then you will have to send him to Italy'.[71]

For nearly two millennia the West rejected zero. 'The consequences were dire. Zero's absence would stunt the growth of mathematics, stifle innovation in science, and, incidentally, make a mess of the calendar.'[72] At the turn of the second millennium, this paucity of European numeracy was made clear to the world when, with much bewilderment, everyone realized that the millennium was not in fact year 2000 but Gregorian year MM. In 1582, when Pope Gregory XIII had introduced modifications into the Julian calendar, he added only a degree of precision, leaving all the essential elements, the eras, the numerals, the festivities and most of all the counting systems unchanged. The Gregorian calendar does not have a year zero, shifting confusingly from 1 BCE to 1 CE. Though it has spread across the planet,[73] and now appears as the dominant time-registry throughout the globe, it operates with the numerical culture of the Roman Empire, which did not include a zero sign.

As the recent puzzle over our calendar clearly reveals, despite the exception of the Gregorian calendar, our numerical culture is decidedly not of the Western type. Not only does the global number system not come from the West, it was actively resisted and even suppressed by Western cultural authority. As John McLeish writes in his book, *Number*, 'So far as the development of calculation is concerned it can not be overstated that the whole line of development lay outside of Europe: in Sumaria, Babylon, China, Indian and the Arabian peninsula. Until the breakthrough by the Arabs in the Scientific renaissance of the 7th to 15th centuries Western Europe was a mathematical backwater. The reason was the baleful legacy of the Greeks.'[74] By tying number to measurement and by fearfully rejecting nothing and the void in favor of the unity of the one 'the story of zero', writes Seife, 'is the story of the Western world's attempt to shield itself unsuccessfully (and sometimes violently) from an Eastern idea'.[75]

Capitalism, zero and the West

At the heart of Joseph Conrad's fictional tale, *The Secret Agent*, is a profound recognition of the centrality of our number system to the capitalist way of life. Set in London in 1884, the key to *The Secret Agent* is related through one vital conversation between the agent, Mr Verloc and his boss, Vladimir, the Russian agent provocateur. Appearing on summons to Vladimir's office, Verloc is informed that in order to earn his keep he must use his links with the anarchist group 'Future of the Proletariat' to oversee a series of terrorist activities designed to arouse extremism among the British, discredit the revolutionaries and provoke panic among the bourgeoisie. In order to accomplish these goals, Vladimir insists, all the traditional modes of terrorism must be abandoned. Attacks on either royalty or religion will not do. Assassinations are expected. Assaults on public buildings, while they undoubtedly cause some alarm, are easily dismissed as the act of a lone maniac. 'A bomb in the National Gallery would', Vladimir concedes, 'make some noise, but not among the right people.' To hit at the heart of the bourgeoisie, Verloc is instructed, one must strike against 'the true fetish of the hour – science and learning'.[76] 'It would be really telling', says his boss, gloating in the persuasiveness of his own logic, 'if one could throw a bomb into pure mathematics.'[77]

Conrad's book sets out to ridicule the cruel and baseless inanity of this unsuccessful gesture. His is a story of the perversion of politics, the evil of ideology, emotional betrayal and personal grief. The notion that Marxist revolutionaries eager to attack global culture should target mathematics appears absurd, presumably even to the most ardent anti-globalizers. Yet there is a profound logic in Conrad's suggestion.

Despite being stubbornly resisted in the past, today the numerals 0 to 9 operate as the semiotics for time, money, science and technology – the key components of capitalism. They are the unrivalled 'language' of modernity and, truncated into the zeroes and ones of the digital computer, have become the underlying plane of cyberspace. In the contemporary world, Hindu-Arab numerals are the nearest thing we have to a global language. 'The Indian system of counting is probably the most successful intellectual innovation ever devised by human beings', writes Barrow. 'It has been universally adopted.'[78]

> There is no Tower of Babel for numbers: once grasped, they are every-where understood in the same way. There are more than four thousand languages, of which several hundred are widespread; there are several dozen alphabets and writing systems to represent them; today, how-ever, there is but one single system for writing numbers. The symbols of this system are a kind of visual Esperanto: Europeans, Asians, Africans, Americans or Oceanians, incapable of communicating by the spoken word, understand each other perfectly when they write numbers using the figures 0, 1, 2, 3, 4 ..., and this is one of the most notable features of our number system.[79]

Yet, oddly, we hear no laments about the homogeneity or 'cultural imperialism' of our number system. There is no widespread clamour to protect local counting practices and little if any pressure to preserve indigenous numeracies even though our number system is much more 'homogenous' than English will ever be. 'Individual cultures speak particular languages; commerce and science increasingly speak English', writes Benjamin Barber, but 'the whole world speaks logarithms and binary mathematics.'[80]

The reasons for this are mysterious. Is it because we are somehow not as invested in our numeracy as we are in our language? Is it because our number system does not come from the West and for that reason is not derided for its universality? Or is it because of a perhaps unconscious sense that the number system operates on a plane of cross-cultural interchange that connects without unifying – that it is not truly 'universal' because it does not begin with unity.

Our number system has functioned for over a thousand years as a network for communication, trade and technological advance. It is a system, however, that is based on relinquishing the importance of the One. The Hindu-Arab numerals were not controlled by any centralized agent or imposed as a unifying force. They spread instead through the practical needs of merchants and traders who established a network of communication independent of any particular territory or State. Through pure pragmatics our number system infected and transformed the channels of commerce and technology. It is only by conforming to the requirement of these channels that 'Western' power grew.

Today zero is central to commerce, navigation, engineering, science and technology, the key elements of contemporary globalization. The fact that these fields rest on a numerical culture that does not start counting from one is suggestive of a diverse communicative cosmos of exchange that refuses subordination to any kind of higher unity. It is the Indian zero – with no core, but only edges – which provides the model for this non-unified network, the true conceptual 'home' of global multiplicity.

New Delhi, September 1999

India Internet World, a conference and trade fair is being held at Pragati Maidan, a vast exhibition center near the heart of downtown. Amongst the dozens of companies who have set up booths promoting Internet portals, software solutions, website and e-biz applications is a start-up called 'wonderonline.com'. What is striking about the company is their advertising slogan. Walk past their booth and you are handed a black and white printout that reads 'spirituality is all about writing smart code'. This is the sign of a culture in process, one which does not rely on assimilation into the homogeneity of a world system, nor on a return to an archaic tradition. It indicates neither a passive nor a reactive periphery, but rather one that actively participates in the creation of a new, innovative global culture that is heterogeneous and unpredictable. To connect spirituality with writing smart code involves an assertion that the local was always – at least virtually – global. For India, and for all it connects to, the future of globalization is decidedly not the Westernization of the world.

Notes

Introduction

1. In October 1998, for example, the state was battered by a 'supercyclone' that killed approximately 10 thousand people.
2. Gurcharan Das, *India Unbound* (New Delhi, New York: Viking, 2000), p. ix.
3. Ibid., p. 346.
4. 'DotKarma' in Cheryl Bentsen and Tom Field, 'India Unbound: CIO Field Report', *CIO: The Magazine for Information Executives*, 1 December 2000, p. 104.
5. 'The Software Industry in India: A Strategic Review' (New Delhi: Nasscom, 1999), p. 3.
6. Ibid., p. 188.
7. The idea of 'modern capitalism', which is discussed in detail in Chapter 1, presupposes a passive periphery, where marginal populations, maladapted to the tumultuous activity of industrial life, were gradually but inexorably absorbed into the alien system of productive rationality. Today the theory has grown to incorporate a more oppositional model of thought, pitting contemporary capitalism against the deep roots of ethnicity, nationalism, religion, and tribal or territorial bonds. The vision of the periphery has become, if not active, at least reactive. No longer innocently malleable, peripheral populations now tend to be seen as clinging to traditions and identities that counter the assimilative forces impinging upon them from outside. In this way they contribute to globalization, but only as its negative – frequently violent – opposition. It is in explicit contrast to these theories that this book proposes a positive conception of the periphery.
8. V.N. Balasubramanyam, *Conversations with Indian Economists* (London: Palgrave, 2001), p. 11. The outrage that this attitude rightfully provokes is forcefully articulated by Shashi Tharoor in a passage that, as Balasubramanyam writes, is worth quoting in full:

[CM Stephen's response was] ignorant (he had no idea of the colossal economic losses caused by poor communications), wrong headed (he saw a practical problem as only an opportunity to score a political point), unconstructive, (responding to complaints by seeking a solution apparently did not occur to him) self righteous (the socialist can't about telephones being a luxury not a right) complacent (taking pride in an eight year waiting list which should have been a sense of shame, since it pointed to the poor performance of his own ministry, unresponsive (feeling no obligation to provide a service in return for the patience and the fees, of the country's telephone subscribers) and insulting (asking long suffering telephone subscribers to return their instruments instead of doing something about their complaints). Ibid.

9. Arvind Singhal and Everett M. Rogers, *India's Communication Revolution: From Bullock Carts to Cyber Marts* (New Dehi: Sage, 2001), p. 57.
10. 'India Versus China: How to Bridge the Gap', *Businessworld*, 10 June 2002.
11. Arvind Singhal and Everett M. Rogers, *India's Communication Revolution: From Bullock Carts to Cyber Marts*, pp. 107–108.
12. 'The Wiring of India', *The Economist*, 25 May 2000.
13. 'One more thing happened during liberalization in the center,' Rekhi continued 'the New Delhi government became very weak, the Congress party, which had ruled for 40 years, was crumbling. The local regional parties were developing... It was at this point that satellite television came in... Before that TV in India was totally in the control of the government. They gave you one message, it was Doordarshan. So as soon as the TV started to beam into India you had these entrepreneurs setting up cable TV... What that did was make the government lose control of information. If you go to India now and listen to the television debates they are very impressive, people are debating at a fundamental level. When I started to go to India I was able to speak on the TV and be heard by 40 million people live without any censorship. That wasn't possible before.' Personal interview with Kanwal Rekhi (Silicon Valley, 17 June 2002).
14. Allen Hammond and Elizabeth Jenkins 'Bottom-up, Digitally Empowered Development', *Information Impacts Magazine*, February 2001.
15. Pramod Mahajan, Paper presented at the Hyderabad IT Forum, Hyderabad (22–24 January 2003).
16. Personal interview with Osama Manzar (New Delhi, 2003).
17. C.K. Prahalad, Paper presented at TiEcon, New Delhi (8 January 2003).
18. C.K. Prahalad, *India as a Source of Innovations: The First Lalbahadur Shastri National Award for Excellence in Public Administration and Management Sciences Lecture*, 30 September, 2000, New Delhi (Digital Dividend, 2000 [cited 2 February 2004]); available from www.digitaldividend.org/pdf/0203ar03.pdf.
19. Tom Field, 'Power Point Prophet', *CIO: The Magazine for Information Executives* (1 December 2000), pp. 158–164.
20. 'What's Stopping Us? Businessworld Round Table', *Businessworld*, 17 February 2003.
21. Personal interview Madanmohan Rao (Bangalore, 18 January 2003).
22. Jagdish Bhagwati, 'Why Your Job isn't Moving to Bangalore', *New York Times*, 15 February 2004.

1   The idea of Westernization

1. Personal interview with Ashish Sen (Bangalore, 20 January 2003).
2. Khushwant Singh quoted in Mehrotra Raja Ram, *Indian English Texts and Interpretations* (Amsterdam, Philadelphia: John Benjamins Publishing Company, 1998), p. 2.
3. R.S. Gupta, 'English in Post-Colonial India', in *Who's Centric Now? The Present State of Post-Colonial Englishes*, ed. Bruce Moore (Oxford: Oxford University Press, 2001), p. 150.

4. Ibid.
5. Braj B. Kachru, *The Alchemy of English: The Spread, Functions, and Models of Non-Native Englishes*, 1st edn, *English in the International Context* (Oxford [Oxfordshire]; New York: Pergamon Institute of English, 1986), p. 1.
6. Ibid., p. 14.
7. Barbara Wallraff, 'What Global Language', *Atlantic Monthly* 2000, p. 53.
8. Robert MacNeil, Robert McCrum and William Crun, *The Story of English* (London: Faber & Faber and BBC Books, 1992), p. 10.
9. Ibid.
10. There are those who argue that the dominance of English – like the so-called universal languages that preceded it (Latin and Greek) – will eventually recede. English's supremacy, these theorists contend, is the result of the cultural and geopolitical power of its native speakers. 'Throughout history,' writes Samuel Huntington, in his immensely influential book *The Clash of Civilizations*, 'the distribution of languages in the world has reflected the distribution of power in the world.' Samuel P. Huntington, *The Clash of Civilizations and the Remaking of World Order*, 1st Touchstone edn (New York: Touchstone, 1997), p. 62.

    English, according to this vision, will lose its prominence at some point in the future when the power of America is inevitably overshadowed by some other state or civilization. 'If at some point in the distant future China displaces the West as the dominant civilization in the world, English', Huntington predicts, 'will give way to Mandarin as the world's lingua franca.' Ibid., p. 63.

    There are various reasons to be suspicious of this claim. Many of them have to do with inherent properties of English that will be discussed in later chapters. Also worth considering, however, is the extreme enthusiasm for English inside China (with both the government and the private sector putting immense resources into teaching English to vast sections of the population) and the fact that many second-generation Chinese living abroad prefer English to their mother tongue, which they find too difficult to learn.
11. Braj B. Kachru, *The Other Tongue: English across Cultures* (Urbana: University of Illinois Press, 1982), p. 3.
12. Thomas de Quincy quoted in Arjya Sircar, 'Indianization of English Language and Literature', in *The Struggle with an Alien Tongue: The Glory and the Grief; Indianisation of English Language and Literature*, ed. R.S. Pathak (New Delhi: Babri Publications), p. 73.
13. Benjamin Barber, *Jihad vs Mcworld: How Globalism and Tribalism are Re-Shaping the World* (J. Ballantine Books, 1996), p. 84.
14. Samir Amin, *Eurocentrism* (New York: Monthly Review Press, 1989), p. 105.
15. Max Weber, 'The Social Psychology of World Religions', in *From Max Weber*, ed. H.H. Gerth and C.W. Mills (New York: Oxford University Press, 1946), p. 269.
16. It would be a mistake to assume that Weber's argument is that the Protestant religion is the cause of capitalism, or the capitalist way of life. 'No economic ethic', he writes, 'has ever been determined solely by religion. In the face of man's attitudes towards the world – as determined by religious or other (in our sense) "inner" factors – an economic ethic has, of course, a high

measure of autonomy. Given factors of economic geography and history determine this measure of autonomy in the highest degree. The religious determination of life-conduct, however, is also one – note this – only one, of the determinants of the economic ethic.' Max Weber, *The Protestant Ethic and the Spirit of Capitalism*, trans. Talcott Parsons (New York: Charles Scribner's Sons, 1958), p. 268.

17. Max Weber, *From Max Weber: Essays in Sociology*, ed. H.H. Gerth and C. Wright Mills (New York: Oxford University Press, 1946), p. 290.
18. Ibid., p. 148.
19. Ibid., p. 289.
20. Weber, *The Protestant Ethic and the Spirit of Capitalism*, pp. 113–114.
21. Weber, *From Max Weber: Essays in Sociology*, p. 290.
22. To quote Weber: 'In inner-wordly asceticism, the grace and the chosen state of the religiously qualified man prove themselves in everyday life. To be sure, they do so not in the everyday life as it is given, but in methodical and rationalized routine-activities of workaday life in the service of the Lord. Rationally raised into a vocation, everyday conduct becomes the locus for proving one's state of grace.' Ibid., p. 291.
23. The three laws of dialectics, the law of the transformation of quantity into quality and vice versa, the law of the interpenetration of opposites and the law of the negation of the negation give, for Marxists, a scientific basis to this succession. Friedrich Engels, 'Dialectics of Nature', in *Reader in Marxist Philosophy*, ed. Howard Selsam and Harry Martel (New York: International Publishers, 1980), pp. 122–123.
24. Karl Marx, *A Contribution to the Critique of Political Economy*, trans. S.W. Ryazanskay (Moscow: Progress Publishers, 1970), p. 21.
25. Joseph Alois Schumpeter, *Ten Great Economists, from Marx to Keynes* (New York: Oxford University Press, 1951), p. 13.
26. Engels, *The Essentials of Marx: The Communist Manifesto*, p. 32.
27. Karl Marx, *Capital*, trans. Eden and Cedar Paul (London: George Allen & Unwin Ltd, 1928), p. 377.
28. Ibid.
29. Ibid.
30. In a letter from Marx to Engels reprinted in Shlomo Avineri (ed.), *Karl Marx on Colonialism and Modernization: His Dispatches and Other Writings on China, India, Mexico, the Middle East and North Africa* (New York: Doubleday, 1968), p. 451.
31. Marx, *Capital*, p. 379. The other main feature in Marx's discussion of the Asiatic mode concerns the state control of public works. This issue is at the heart of the notion of 'Oriental Despotism' and is taken up, in great detail, by the work of Karl Wittfogel.
32. One may argue that Marx is here providing a materialist basis for a Hegelian prejudice. To quote Hegel: 'On the one side we see duration, stability – Empires belonging to mere space, as it were [as distinguished from Time] – unhistorical History; ... the States in question, without undergoing any change in themselves, or in principle of their existence are constantly changing their position towards each other. They are in ceaseless conflict, which brings rapid destruction. This history too is, for the most

part, really *unhistorical*, for it is only the repetition of the same majestic ruin.' Georg Hegel, *Philosophy of History* (New York: Dover Publications, 1956), pp. 105–106.

33. Karl Marx, 'The Future Results of British Rule in India', *New York Daily Tribune*, 8 August 1853. This notion of the ahistorical character of the Asiatic mode of production gives an odd complexity to Marx's analyses of colonialism. Though, on the one hand, Marx deplored the universalizing tendencies of capitalism, the principles of historical materialism were such that he was forced to see colonialism as a necessary evil. Capitalism could only change internally by subsuming the entire globe. Thus, English commerce, according to Marx, exerted a 'revolutionary influence' on Indian communities. By enforcing change at the level of production colonialism brought with it the first *social* revolution that had ever occurred on Indian soil, inserting Indian society, for the first time, into the dynamism of historical change. To quote Marx:

> Now sickening as it must be to human feeling to witness these myriads of industrious patriarchal and inoffensive social organizations disorganized and dissolved into their units, thrown into a sea of woes, and their individual members losing at the same time their ancient form of civilization and their hereditary means of sustenance, we must not forget that these idyllic village communities, inoffensive though they may appear, had always been the solid foundation of Oriental despotism, that they restrained the human mind within the smallest possible compass, making it the unresisting tool of superstition, enslaving it beneath traditional rules, depriving it of all grandeur and historical energies...We must not forget that this undignified, stagnatory, and vegetative life, that this passive sort of existence, evoked on the other part, in contradistinction, wild, aimless, unbounded forces of destruction, has rendered murder itself a religious rite in Hindustan. We must not forget that these little communities were contaminated by distinctions of caste and by slavery, that they subjugated man to external circumstances instead of elevating man to be the sovereign of circumstances, that they transformed a self-developing social state into never changing destiny...England it is true, in causing a social revolution in Hindustan, was actuated only by the vilest of interests, and was stupid in her manner of enforcing them. But that is not the question. The question is, can mankind fulfill its destiny without a fundamental revolution in the social state of Asia? If not, whatever may have been the crimes of England she was the unconscious tool of history in bringing about the revolution. Marx, The Future Results of British Rule in India, *New York Daily Tribune*, 8 August 1853.

34. Engels, *The Essentials of Marx: The Communist Manifesto*, p. 34.
35. Ibid., p. 36.
36. Karl Marx, *Grundrisse: Foundations of the Critique of Political Economy* (New York: Vintage Books, 1973), p. 646.
37. Marx, *Capital*, p. 863.
38. Baran A. Paul and E.J. Hobsbawm, 'The Stages of Economic Growth: A Review', in *The Political Economy of Development and Underdevelopment*, ed. Charles K. Wilber (New York: Random House, 1973), p. 9.

39. Terence Hopkins *et al.*, 'Patterns of Development of the Modern World-System', in *World-Systems Analysis: Theory and Methodology*, ed. Immanuel Wallerstein and Terence K. Hopkins (London: Sage Publications, 1982), p. 279.
40. Amin, *Eurocentrism*, p. 71.
41. The quote comes from Immanuel Maurice Wallerstein, *The Capitalist World-Economy: Essays, Studies in Modern Capitalism* (Cambridge [England]; New York: Cambridge University Press, 1979), p. 19. The idea, however, is inherited directly from Marx. 'The discovery of gold and silver in America, the extirpation, enslavement and entombment in mines of the aboriginal population, the beginning of the conquest and looting of the East Indies, the turning of Africa into a warren for the commercial hunting of black skins', wrote Marx in a famous passage, 'signals the rosy dawn of the era of capitalist production.' Marx, *Capital*, p. 832.
42. Terence K. Hopkins and Immanuel Maurice Wallerstein, *World-Systems Analysis: Theory and Methodology, Explorations in the World-Economy; V. 1* (Beverly Hills, Calif.: Sage Publications, 1982), p. 47.
43. 'No one contests the self-evident fact that worldwide capitalist expansion has been accompanied by a flagrant inequality among its partners. But are these the result of a series of accidents due for the most part to various detrimental internal factors that have slowed the process of "catching up?" Or is this inequality the product of capitalist expansion itself and impossible to surpass within the framework of this system?' Amin, *Eurocentrism*, p. 109.
44. Ibid., p. 107.
45. Ibid., p. 75.
46. Gurcharan Das, *India Unbound* (New Delhi, New York: Viking, 2000), p. 74.
47. Terence K. Hopkins, 'The Study of the Capitalist World Economy: Some Introductory Considerations', in *World Systems Analyses: Theory and Methodology*, ed. Terence, K. Hopkins and Immanuel Maurice Wallerstein, *Explorations in the World-Economy; V. 1* (Sage Publications, 1982), p. 13.
48. Daniel Yergin and Joseph Stanislaw, *The Commanding Heights: The Battle for the World Economy* (New York: Simon & Schuster), p. 71.
49. Das, *India Unbound*, p. 86.
50. Ibid., p. 126.
51. Daniel Yergin and Joseph Stanislaw, *The Commanding Heights: The Battle for the World Economy*, Rev. and updated edn (New York: Simon & Schuster, 2002), p. 53.
52. Ibid., p. 59.
53. Jawaharlal Nehru quoted in Das, *India Unbound*, p. 170.
54. Ibid., p. 90.
55. The phrase comes from Lenin, who coined it in November 1922 in a speech made at the Fourth Congress of the Communist International in St Petersburg. Daniel Yergin and Joseph Stanislaw, *The Commanding Heights: The Battle for the World Economy*, p. xii.
56. Das, *India Unbound*, p. 157.
57. Ibid., p. 158.
58. Daniel Yergin and Joseph Stanislaw, *The Commanding Heights: The Battle for the World Economy*, p. 90.

2   Opening up

1. Susan Sontag, 'The World as India', *Times Literary Supplement*, 6 June 2003. Accessed through the online archive at http://www.the-tls.co.uk/.
2. Ibid.
3. Ibid.
4. 'Outsourcing to India: Back Office to the World', *Economist*, 5 May 2001.
5. Ibid.
6. Apratim Barua, 'Cyber-Coolies, Hindi and English; Letter', *Times Literary Supplement* 2003. Accessed through the online archive at http://www.the-tls.co.uk/.
7. Harish Trivedi, 'Cyber-Coolies, Hindi and English; Letter', *Times Literary Supplement*, 27 June 2003. Accessed through the online archive at http://www.the-tls.co.uk/.
8. Ibid.
9. Ibid.
10. Harish Trivedi, 'Cyber-Coolies, Hindi and English', *Times Literary Supplement*, 22 August 2003. Accessed through the online archive at http://www.the-tls.co.uk/.
11. Trivedi, 'Cyber-Coolies, Hindi and English; Letter'.
12. Barua, 'Cyber-Coolies, Hindi and English; Letter'.
13. Ibid.
14. Ibid.
15. Ibid.
16. Trivedi, 'Cyber-Coolies, Hindi and English'.
17. Ibid.
18. Ibid.
19. Ibid.
20. Gurcharan Das, 'Cyber-Coolies, Hindi and English; Letter', *Times Literary Supplement*, 12 September 2003. Accessed through the online archive at http://www.the-tls.co.uk/.
21. Ibid.
22. Ibid.
23. Ibid.
24. Joshua A. Fishman, 'The New Linguistic Order', in *Globalization and the Challenges of a New Century: A Reader*, ed. Howard D. Mehlinger, Patrick O'Meara and Mathew Krain (Bloomington: Indiana University Press, 2000), p. 436.
25. Braj B. Kachru, *The Indianization of English: The English Language in India* (Delhi; New York: Oxford, 1983), p. 19.
26. R.S. Gupta and Kapil Kapoor (eds), *English in India Issues and Problems* (Academic Foundation, 1991), p. 15.
27. Ibid., p. 36.
28. Braj B. Kachru, *The Alchemy of English: The Spread, Functions, and Models of Non-Native Englishes*, 1st edn, *English in the International Context* (Oxford [Oxfordshire]; New York: Pergamon Institute of English, 1986), p. 7.
29. The colonial belief in the superiority of English over any of India's indigenous tongues is summed up by Macaulay's oft repeated statement that 'a single shelf of a good European library was worth the whole native literature of India and Arabia'.

30. Ibid., p. 6.
31. This view was held explicitly from at least as early as 1792, when Charles Grant prepared the first formal blue print on language and education in India. His treatise was entitled 'the State of Society among the Asiatic Subjects of Great Britain, particularly with respect to Morals and the Means of Improving It'. This blue print called for English to 'be introduced in India as the medium of instruction in a Western system of education'. It also suggested that 'English be adopted as the official language of the Government for easy communication between the rulers and the ruled'.

    Grant outlined three main objectives: '(1) Religious: Wherever this knowledge would be received, idolatry with all the rabble of its impure deities, its monsters of wood and stone . . . would fall. The reasonable service to the only and infinitely perfect God would be established. (2) Commercial: In every progressive step of this work i.e. education and conversion, we shall also serve the original design with which we visited India, that design still so important to this country – the extension of our commerce. (3) Political: Through education, the Indians would be brought nearer to the rulers. The teaching of the Gospel would ensure that they remain loyal.' Gupta (ed.), *English in India Issues and Problems*, p. 31.
32. Ibid., p. 32.
33. Ibid., p. 46.
34. Archana S. Burde and N. Krishnaswamy, *The Politics of Indians English* (Oxford: Oxford University Press, 1998), p. 13.
35. Arjya Sircar, 'Indianization of English Language and Literature', in *The Struggle with an Alien Tongue: The Glory and the Grief: Indianisation of English Language and Literature*, ed. R.S. Pathak (New Delhi: Babri Publications), p. 64.
36. 'Non native speakers', writes Kachru, 'must avoid regarding English as an evil influence which necessarily leads to Westernization. In South Asia and South Africa the role of English in developing nationalism and mobilizing the intelligentsia at large for struggles toward freedom cannot be over-emphasized.' Braj B. Kachru, *The Other Tongue: English across Cultures* (Urbana: University of Illinois Press, 1982), p. 51.
37. Robert MacNeil, Robert McCrum and William Crun, *The Story of English* (London: Faber & Faber and BBC Books, 1992), p. 33.
38. William Richter, 'The Politics of Language in India' (PhD Dissertation, University of Chicago, 1968), p. 49.
39. Gandhi and Nehru both sought to replace English with Hindustani – a language that combines Hindi and Urdu and could thus be seen as a unifying force between Hindus and Muslims – as the indigenous tongue that was best suited to become India's national language. In 1925 they adopted Hindustani for the official proceedings of Congress.
40. Mehrotra Raja Ram, *Indian English Texts and Interpretations* (Amsterdam, Philadelphia: John Benjamins Publishing Company, 1998), p. 5.
41. Gandhi quoted in Tulsi Ram, *Trading in Language: The Story of English in India* (New Delhi: GDK Publications, 1983), p. 5.
42. David Crystal, *English as a Global Language* (Cambridge, England; New York: Cambridge University Press, 1997), p. 114.
43. Gupta and Kapil Kapoor (eds) *English in India Issues and Problems*, p. 130.

44. Ram, *Indian English Texts and Interpretations*, p. 6.
45. Richter, 'The Politics of Language in India', p. 28.
46. Ram, *Indian English Texts and Interpretations*, p. 1.
47. Ram, *Trading in Language: The Story of English in India*, p. 275.
48. Jyotindra Dasgupta, *Language Conflict and National Development: Group Politics and National Language Policy in India* (Berkeley: University of Berkeley Press, 1970), p. 162.
49. Among the attacks on Hindi none were so vehement as those launched by the supporters of English. Frank Anthony, an eminent Anglo-Indian leader, who attended the major language conferences in the South, repeatedly warned against the new Hindi. He is quoted as saying that 'the new Hindi is a symbol of all that is reactionary and retrograde in the country. The new Hindi today is the symbol of communalism; it is a symbol of religion; it is the symbol of language chauvinism, and worst of all, it is the symbol of oppression of the minority languages.' Ibid., p. 193.
50. Ibid., p. 192.
51. Ibid.
52. Ibid., p. 227.
53. Ibid., p. 237.
54. MacNeil, *The Story of English*, p. 367.
55. Gurcharan Das, *India Unbound* (New Delhi; New York: Viking, 2000), p. 98.
56. Yergin, *The Commanding Heights: The Battle for the World Economy*, pp. 214–215.
57. Ibid.
58. Das, *India Unbound*, p. 94.
59. In an interview, Kailash Joshi described the process in the following terms: The existing industry in India, the large industry houses were doing OK. There were big markets so they would make whatever junk product they make, sell it and they were very happy. Then they had the politicians and then they had the bureaucrats. These three said 'hey we've got a good thing going here'. 'Why don't we limit this entrepreneurship to ourselves? Let's have fun here. You guys you keep selling the junk you make. Give us enough money for our elections and the bureaucrats – well there was some corruption there. And so they took this pie and they kept going around in a circle... (Personal interview with Kailash Joshi, Silicon Valley, 17 June 2003).
60. Statistics are from the World Bank website available at http://www.worldbank.org/data.
61. Sonia Gandhi resisted recruitment into the Congress party until 1997. She became party president in 1998.
62. Das, *India Unbound*, p. 214.
63. Ibid., p. 213.
64. Ibid., p. 215.
65. Singh quoted in Yergin, *The Commanding Heights: The Battle for the World Economy*, p. 221.
66. Ibid., p. 219.
67. Das, *India Unbound*, p. 216.
68. Ibid., p. 219.
69. 'A Celebration of Freedom', *Businessworld*, 10 January 2000, p. 70.
70. Archana Burde and Krishnaswamy, *The Politics of Indians' English*, p. 38.
71. Ram, *Trading in Language: The Story of English in India*, p. 48.

72. V.N. Balasubramanyam, *Conversations with Indian Economists* (London: Palgrave, 2001), p. 49.
73. V.K. Gokak, *English in India: Its Present and Future* (New York: Asia Publishing House, 1964), p. 5.
74. Nehru quoted in Richter, 'The Politics of Language in India', p. 133.
75. MacNeil, *The Story of English*, p. 33.
76. Ram, *Indian English Texts and Interpretations*, p. 8.
77. Unicode – a coding scheme that can handle most of the world's diverse languages is beginning to replace ASCII and is becoming a universal standard.
78. Traditional ASCII was built as a 7-bit code, and thus could only represent 128 characters. Extended ASCII is an 8-bit system with a possibility to represent 256 characters and some built-in flexibility.
79. From the British Council Frequently Asked Questions website http://www.britishcouncil.org/english/engfaqs.htm.
80. Statistics are from Global Reach available at http://www.glreach.com/globstats/index.php3.
81. Statistics are from Global Reach available at http://global-reach.biz/globstats/refs.php3.

3 Eastern influences

1. Hi-Tec city is an acronym for Hyderabad Information Technology Engineering and Consultancy City. It is a joint venture between Larsen & Toubro Limited (L&T) which has 80 percent equity and the Andhra Pradesh Industrial Infrastructure Corporation Ltd. Hi-Tec city is described in the company's literature as an 'ultra modern techno township'. It will evolve over a 12-year period. The plan calls ultimately for 5 million square feet of built-up area with facilities for electronics, computer software, hardware, engineering and consultancy, banking and financial services.
2. Mircea Eliade, *Yoga: Immortality and Freedom* (Princeton: Bollingen, 1969), p. 19.
3. Sam Pitroda, Paper presented at the TiEcon, New Delhi (8 January 2003).
4. Rajat Gupta, Paper presented at the TiEcon, New Delhi (8 January 2003).
5. 'What's Stopping Us? Businessworld Round Table', *Businessworld*, 17 February 2003.
6. C.K. Prahalad, Paper presented at the TiEcon, New Delhi (8 January 2003).
7. By early 2004 there were signs that this was beginning to occur. In January 2004 *The Economist* reported that India's GDP was growing at 8.4 percent 'Emerging-Market Indicators', *The Economist*, 17 January 2004.
8. Daniel Yergin and Joseph Stanislaw, *The Commanding Heights: The Battle for the World Economy* (Rev. and updated edn) (New York: Simon & Schuster, 2002), p. 220.
9. This is most evident with the Internet, which introduces a new contemporaneity to transcultural interchanges, substituting horizontal diffusion ('connectivity') for vertical or historical comparability.
10. 'Wet markets' is a name for the lively street bazaars in Asia that sell fresh – and often live – food.
11. Tharman Shanmugaratnam, Paper presented at the Hyderabad IT Forum, Hyderabad (22–24 January 2003).

12. To quote from *The Economist* magazine: 'For decades, people talked of Asia's economic miracle when what they really meant was East Asia's.' 'Let It Shine', *The Economist*, 21 February 2004.

13. Patwardhan continues: 'The amount of familiarity that the Indian intelligentsia have with the West versus their own neighboring countries sometimes really boggles the mind. I mean how could we have let it go on for so long? I find it absurd that I don't have a daily flight to Bangkok – here's this country with whom we've had religious and cultural interactions for thousands of years – and it's literally as far away from Bombay as Assam and I don't have daily flight, yet I have half a dozen daily flights to cities in Europe.' From a personal interview with Anand Patwardhan (Mumbai, 6 February 2003).

14. Ibid.

15. Personal Interview with Kanwal Rekhi (Silicon Valley, 17 June 2002).

16. Personal Interview with Madanmohan Rao (Bangalore, 18 January 2003).

17. Ibid.

18. Personal Interview with Kanwal Rekhi.

19. 'What's Stopping Us? Businessworld Round Table', *Businessworld*, 17 February 2003.

20. V.N. Balasubramanyam, *Conversations with Indian Economists* (London: Palgrave, 2001), p. 127.

21. Jean Dreze and Amartya Sen, *India Development and Participation* (New Delhi: Oxford University Press, 2002), p. 114.

22. Gurcharan Das, *India Unbound* (New Delhi; New York: Viking, 2000), p. 25.

23. Jean Dreze and Amartya Sen, *India Development and Participation*, p. 115.

24. Statistics taken from 'India Versus China: How to Bridge the Gap', *Businessworld*, 10 June 2002 and a report by i-Watch an Indian NGO 'Wake up India' (Mumbai: i-watch, 2002), www.india-watch.com.

25. It remains unclear whether India's 2004 figure of above 8 percent growth signals the beginning of a long-term trend.

26. Joanna Slater, 'In the Zone', *Far Eastern Economic Review*, 8 May 2003.

27. Niranjan Rajadhyaksha, 'Culture Matters', *Businessworld*, 19 March 2001.

28. Pitroda.

29. Niranjan Rajadhyaksha, 'The Chinese Opportunity', *Businessworld*, 17 February 2003.

30. Ibid.

31. 'China Crises', *The Economist*, 7 June 2003.

32. Personal interview with Sunil Mehta (New Delhi, 13 January 2003).

33. Das, *India Unbound*, p. 25.

34. Yergin, *The Commanding Heights: The Battle for the World Economy*, p. 213.

35. Jean Dreze and Amartya Sen, *India Development and Participation*, p. 142.

36. R.S. Shah, Paper presented at the TiEcon, New Delhi (8 January 2003).

37. Gurcharan Das, *The Elephant Paradigm: India Wrestles with Change* (New Delhi: Penguin, 2002), p. ix.

38. 'What's Stopping Us? Businessworld Round Table'.

39. Tavleen Singh, 'A Freedom Foiled', *India Today*, 27 May 2002.

40. Das, *India Unbound*, p. 213.

41. Ibid., p. 224.

42. Ibid., p. 213.

43. If any political party benefits from reforms it is the BJP, since they have been in power during the time reforms had their most palpable effect on people's daily life.
44. Ibid., p. 213.
45. Chandrababu Naidu's loss in the 2004 election further intensifies questions about the relationship between democracy and economic growth. For more on this, see Anna Greenspan, *Taxed by Democracy?* in Tech Central Station (2 June 2004). Available at http://www.techcentralstation.com/advertising-sky.htm.
46. Editorial, 'Dr Politics and Mr Economics', *Businessworld*, 4 September 2000, p. 72.
47. Erla Zwingle, 'World Cities', *National Geographic*, November 2002.
48. Ibid., p. 95.
49. Ibid.
50. Ibid.
51. Personal communication.
52. Naidu quoted in *Businessworld*, 25 October 1999.
53. 'A visit would be worth the small handling fee if only for the blast of air conditioning,' continues Zwingle writing for the *National Geographic*. 'The office was full. In the first two hours of business, tellers at all the centers handled 1.4 million rs of transactions and they were going to be open for another eight hours.' Zwingle, 'World Cities'.
54. Ibid.
55. This topic is discussed in detail in Chapter 7.

4 Marginal capitalisms

1. TiE now uses a more open less ethnically centered motto: Talent Ideas Enterprise, which will discussed in more detail in Chapter 8.
2. AnnaLee Saxenian, *Silicon Valley's New Immigrant Experience* (California: Public Policy Institute of California, 1999), p. 10.
3. Anthony Spaeth, 'The Golden Diaspora', *Time*, 19 June 2000.
4. Saxenian, *Silicon Valley's New Immigrant Experience*, p. 10.
5. H1-B visa's are granted for specialty jobs that cannot be filled by local applicants.
6. 'According to the INS, nearly 50 per cent of the H1-B professionals are Indian and at least half of these jobs were related to programming and systems analysis. The incoming influx of H1-B workers from India has boosted the total Indian population to 1.7 million.' Lavina Melwani, *Back2bangalore* (www.littleindia.com, 2001 [cited 9 March 2004]); available from http://www.littleindia.com/India/July2001/B2B.htm.
7. 'The Indian population over the past decade more than doubled in other tech hubs: Fairfax County, Va.; Middlesex County, Mass.; and King County, Wash., headquarters of Microsoft.' Rachel Konrad, *Chasing the Dream* (News.Com, 2001 [cited 3 February 2004]); available from http://news.com.com/ 2009-1017_3-270647.html.
8. Saxenian, *Silicon Valley's New Immigrant Experience*, p. v.
9. Ibid., p. viii.
10. 'E-Billionaires', *New Indian Express*, 18 September 1999.
11. Vidya Viswanathan, 'Indian Internet Mafia', *Businessworld*, 24 May 1999.

12. Spaeth, 'The Golden Diaspora'.
13. *Angel Investor News* defines an angel investor in their glossary as 'individuals who invest in businesses looking for a higher return than they would see from more traditional investments. In return for their investment they often are highly involved in the business. Usually they are the bridge from the self-funded stage of the business to the point that the business needs the level of funding that a venture capitalist would offer.' From http://www.angel-investor-news.com/glossary.htm.
14. Max Weber, *Max Weber on Capitalism, Bureaucracy and Religion*, ed. Stanislav Andreski (London: George Allen & Unwin, 1983), p. 158.
15. Ibid., p. 115.
16. Max Weber, *The Protestant Ethic and the Spirit of Capitalism*, trans. Talcott Parsons (New York: Charles Scribner's Sons, 1958), p. 182.
17. Karl Marx and Frederick Engels, *The Essentials of Marx: The Communist Manifesto* (New York: Vanguard Press, 1926), p. 31.
18. Max Weber, *From Max Weber: Essays in Sociology*, eds H.H. Gerth and C. Wright Mills (New York: Oxford University Press, 1946), p. 216.
19. Ibid., p. 228.
20. Ibid., p. 214.
21. Joseph Alois Schumpeter, *Capitalism, Socialism, and Democracy*, 2nd edn (New York, London: Harper & Brothers, 1947), p. 206.
22. Weber, *Max Weber on Capitalism, Bureaucracy and Religion*, p. 159.
23. Schumpeter, *Capitalism, Socialism, and Democracy*, p. 206.
24. Karl Marx, *Capital*, trans. Eden and Cedar Paul (London: George Allen & Unwin Ltd, 1928), pp. 690–691.
25. Weber, *From Max Weber: Essays in Sociology*, p. 215.
26. Schumpeter, *Capitalism, Socialism, and Democracy*, p. 83.
27. Ibid., p. 134.
28. Ibid., p. 133.
29. Ibid.
30. Ibid., p. 132.
31. Ibid., p. 133.
32. Kotkin, *Tribes: How Race, Religion, and Identity Determine Success in the New Global Economy*, p. 3.
33. Ibid.
34. This vision has come to dominate thinking about the contemporary relation between culture and capitalism, structuring almost all our presuppositions according to an oppositional duality. On one pole is the emptiness of capitalist culture, a Westernized, homogeneous, shallow McWorld. On the other there exists a kind of cultural backlash in which traditional – and often clashing – national, ethnic or religious identities are strengthened and affirmed.
35. Kotkin, *Tribes: How Race, Religion, and Identity Determine Success in the New Global Economy*, p. 3.
36. Constance Lever-Tracy, David Fu-Keung Ip and Noel Tracy, *The Chinese Diaspora and Mainland China: An Emerging Economic Synergy* (Houndmills, Basingstoke, Hampshire, New York: Macmillan Press; St Martins Press, 1996), pp. 24–25.
37. Ivan Light and Steven J. Gold, *Ethnic Economies* (San Diego: Academic Press, 2000), p. 7.

38. Lever-Tracy, Ip and Tracy, *The Chinese Diaspora and Mainland China: An Emerging Economic Synergy*, pp. 24–25.
39. Light, *Ethnic Economies*, p. 6.
40. Lever-Tracy, Ip and Tracy, *The Chinese Diaspora and Mainland China: An Emerging Economic Synergy*, p. 12.
41. Kotkin, *Tribes: How Race, Religion, and Identity Determine Success in the New Global Economy*, p. 4.
42. Light, *Ethnic Economies*, p. 110.
43. Ivan Light, 'Immigrant and Ethnic Enterprise in North America', *Ethnic and Racial Studies* 7: 2 (1984), p. 199.
44. Personal Interview with Kanwal Rekhi (Silicon Valley, 17 June 2002).
45. Lever-Tracy, Ip and Tracy, *The Chinese Diaspora and Mainland China: An Emerging Economic Synergy*, p. 8.
46. The largest is the African diaspora.
47. Lever-Tracy, Ip and Tracy, *The Chinese Diaspora and Mainland China: An Emerging Economic Synergy*, p. 15.
48. Ibid., p. 21.
49. Light, *Ethnic Economies*, p. 95.
50. Lever-Tracy, Ip and Tracy, *The Chinese Diaspora and Mainland China: An Emerging Economic Synergy*, p. 23.
51. Kotkin, *Tribes: How Race, Religion, and Identity Determine Success in the New Global Economy*, p. 204.
52. Ibid., p. 9.
53. Ibid., p. 232.
54. Warner, 'The Indians of Silicon Valley'.
55. The conference itself required the work of more than 300 volunteers for over 6 months.
56. Personal interview with Kailash Joshi (Silicon Valley, 17 June 2002).
57. Personal interview with Kanwal Rekhi.
58. Ibid.
59. Ibid.
60. Personal interview with Vish Mishra (Silicon Valley, 17 June 2002).
61. Personal interview with Kanwal Rekhi.
62. Ibid.
63. Ibid.
64. Ibid.
65. Ibid.
66. Personal interview with Kailash Joshi. Though TiE's origin in Silicon Valley ensures its tight connection with the high-tech industries, this is not an essential aspect of the organization. Chapters in other regions – where the tech sector is not so important – have developed to serve the local economy. In Detroit, for example, TiE is active in the automative industry; TiE's chapter in Florida has links with the tourism industry; and in Jaipur, India, it has a special interest group on gems, jewelry and handicrafts. This diversification does not pose a problem since it does not stray from TiE's core mission. As Mishra says 'so long as you keep fostering entrepreneurship that is the key'. Personal interview with Vish Mishra.
67. Personal interview with Kanwal Rekhi.
68. Ibid.

69. Personal interview with Vish Mishra.
70. Ibid.
71. Personal interview with Kanwal Rekhi.
72. 'Dot Hatao, Desh Bachao', *Economic Times Online*, 2000.
73. All networks, whether cultural, business or technological, can be characterized by certain key features:

- A network is at once both single and multiple. 'No other arrangement,' writes Kevin Kelly, 'chain, pyramid, tree, circle, hub – can contain true diversity working as a whole.' Networks operate as open platforms, which seek to facilitate diverse flows of information, communication and trade. Though integrated, they are not at all homogenous. 'The Net is an emblem of multiples.' Completely interconnected it is not at all 'the same', since each node is free to connect with the whole in a myriad of ways.
- Networks are flat systems that cannot be controlled from above. True networks are rigidly nonhierarchical, growing from 'the bottom-up'. Though they may comply with standards or rules, these are typically protocols of information rather than instructions to be obeyed. In a network no superior power is in a position to dictate commands. This is why Kelly writes that 'hidden in the Net is the mystery of the Invisible Hand – control without authority'.
- Networks constitute intrinsically decentralized systems, whose behavior cannot be determined or predicted by reference to a command core or 'central planning' authority. 'The Net icon has no center – it is a bunch of dots connected to other dots – a cobweb of arrows pouring into each other, squirming together like a nest of snakes, the restless image fading at indeterminate edges.' A true network consists entirely of periphery.

For more on networks and network science see: Kevin Kelly, *Out of Control: The New Biology of Machines, Social Systems and the Economic World* (Cambridge: Perseus Books, 1994). Available online at http://www.kk.org/outofcontrol/contents.php and Albert-Laszlo Barabasi, *Linked: The New Science of Networks* (Perseus, 2002).

74. Lever-Tracy, Ip and Tracy, *The Chinese Diaspora and Mainland China: An Emerging Economic Synergy*, p. 13.
75. For more on this see 'The Spacemen Have Landed' in Joel Kotkin, *Tribes: How Race, Religion and Identity Determine Success in the New Global Economy*.
76. Lever-Tracy, Ip and Tracy, *The Chinese Diaspora and Mainland China: An Emerging Economic Synergy*, p. 13.
77. Ibid., p. 27.
78. Ibid., p. 31.
79. Ibid., p. 6.
80. Ibid., p. 11.
81. Ibid., p. 86.
82. Ibid., p. 282.
83. Ibid., p. 280.

84. Satyajit Datta, 'Look, Here's a Booming Market!', *Businessworld*, 7 February 1999.
85. 'Magnates in Manacles', *Businessworld*, 21 May 1999.
86. Ibid.
87. Saxenian, *Silicon Valley's New Immigrant Experience*, p. 63.
88. Cheryl Bentsen and Tom Field, 'India Unbound: CIO Field Report', *CIO: The Magazine for Information Executives*, 1 December 2000, p. 102.
89. Personal Interview with Kailash Joshi.
90. This is discussed in greater detail in Chapter 7.
91. Vinod Khosla, Paper presented at the TiEcon, Silicon Valley (2002).
92. http://www.aifoundation.org/.
93. http://www.ashanet.org/.
94. http://www.ffe.org/.
95. Personal interview with Kailash Joshi.
96. Ibid.
97. Ibid.
98. Ibid.
99. Kanwal Rekhi, Paper presented at the TiEcon, Silicon Valley (14–15 June 2002).
100. Personal interview with Kanwal Rekhi.
101. Ibid.
102. Viswanathan, 'Indian Internet Mafia'.
103. Personal interview with Kanwal Rekhi.
104. AnnaLee Saxenian, 'The Bangalore Boom: From Brain Drain to Brain Circulation?', eds Kenneth Kenniston and Deepak Kumar (Bangalore: National Institute of Advanced Study, 2000).
105. AnnaLee Saxenian, 'Local and Global Networks of Immigrant Professionals in Silicon Valley' (Public Policy Institute of California, 2002), p. 51.

5 The technological edge

1. The only possible competition is Mumbai. Yet, as Ashok Desai writes, 'if Bombay was India's most cosmopolitan city a quarter century ago, it is Bangalore today'. Ashok Desai, 'Why Bangalore Won', *Businessworld*, 23 September 2002, p. 10.
2. Ibid. On 6 January 2004 Gagan Gupta reported in a piece entitled *Bangalore overtakes Silicon Valley* that Bangalore now has 150,000 engineers (http://www.techtree.com/techtree/jsp/showstory.jsp?storyid = 4325).
3. 'Bangalore India is just like here, [it has the] same climate and that has to do with diversity ... This area – Silicon Valley ... what are its distinguishing features? This area has been very welcoming of ideas and people. Bangalore has the same thing they are very open. Other places in India – like Madras (Chennai) – are very closed minded they don't like other people to come in. I studied in Bangalore as an 18-year-old from way up North, they were so welcoming I felt I was at home even though I didn't speak the language. I said this is a great place. People are so friendly. Then on my way home I would take a train and go to Chennai and sometimes wait five hours. They would look at me this guy doesn't speak Tamil, something is wrong with him. So Chennai did not progress as much. I think how you treat humans, and how hospitable you are. I think it has a lot to do with

prosperity in the area.' Personal interview with Kailash Joshi (Silicon Valley, 17 June 2002).

4. Bangalore's population is projected to reach 7 million by 2010. Meenu Shekar, 'The Changing Face of Bangalore', *Businessworld*, 22 July 1997.

5. Tom Field 'For a Few Rupees More' in Cheryl Bentsen and Tom Field, 'India Unbound: CIO Field Report', *CIO: The Magazine for Information Executives*, 1 December 2000, p. 168.

6. From the 1999 Nasscom–McKinsey report.

7. Bentsen, 'India Unbound: CIO Field Report', p. 174.

8. Mehta died suddenly in 2001.

9. 'Power Lobbying', *Business India*, 19 February to 4 March 2001, p. 52.

10. 'Q&A with Ajit Balakrishnan, Chairman, Founder and CEO of Rediff on the Net', *Asia Source*, 13 September 1999.

11. Personal interview with Madanmohan Rao (Bangalore, 18 February 2003).

12. Personal interview with Manas Patnaik (Bhubaneshwar, 1999).

13. Gurcharan Das, *The Elephant Paradigm: India Wrestles with Change* (New Delhi: Penguin, 2002), p. 61.

14. Mumbai is the one city in India where there is very good electricity supply. This is because it is being managed by a private company – Tata Electricity.

15. Bentsen, 'India Unbound: CIO Field Report', p. 183.

16. Asma Lateef, 'Linking up with the Global Economy: A Case Study of the Bangalore Software Industry', *International Institute for Labour Studies* (1996–1997), p. 7.

17. See Edward Yourdon, *Decline & Fall of the American Programmer, Yourdon Press Computing Series* (Englewood Cliffs, NJ: Yourdon Press, 1992).

18. The first Indian Institute of Technology was established in May 1950 in Kharagpur, West Bengal at the site of Hijli Detention camp. Four other campuses were subsequently founded at Mumbai (1958), Chennai (1959), Kanpur (1960) and New Delhi (1961). In 1995, a sixth campus at Guwahati was added and most recently in 2001, a seventh campus was established by upgrading Roorkee University, one of India's oldest engineering institutions, into an IIT.

19. Kamla Bhatt, 'IITs', *Outlook*, 3 February 2003.

20. Ibid.

21. Ibid.

22. Personal interview with Anand Patwardhan (Mumbai, 6 February 2003).

23. http://www.niit.com/niit/home.asp.

24. www.iiit.net.

25. 'Growth Scenarios of IT Industries in India', *Communications of the ACM* 44: 7 (2001).

26. Lateef, 'Linking up with the Global Economy: A Case Study of the Bangalore Software Industry', p. 8.

27. Ibid., p. 9.

28. 'Interview with Dewang Mehta', *Financial Times*, 1999.

29. 'Growth Scenarios of IT Industries in India'.

30. www.tcs.com.

31. www.wipro.com.

32. 'Ever since 1991', Murthy is quoted as saying, 'there has not been a single instance when I went to Delhi for any license for any business of Infosys.

Today I can import a computer worth millions of dollar without having to see a single bureaucrat or apply for a license.' Daniel Yergin and Joseph Stanislaw, *The Commanding Heights: The Battle for the World Economy* (Rev. and updated edn) (New York: Simon & Schuster, 2002), p. 228.

33. Clay Chandler, 'Asia's Businessmen of the Year: They Get IT', *Fortune*, 17 February 2003, p. 41.
34. See Alam Srinivas, 'A Debugged Operating $ystem', *Outlook*, 27 January 2003.
35. Since 19 is the systems cipher the year 00 is treated as 1900 and not 2000.
36. Note the following 1997 quote from *The Economist* magazine: 'Care for a thrill? Consider what might happen if the Millennium Bug, that tendency for many of the world's computers to mistake the year 2000 for 1900, is not eradicated in time...Could two measly digits really halt civilization?' 'Yes, Yes – 2000 Times Yes!', *The Economist*, 4–10 October 1997.
37. 'The World in 1998', *Economist Publications* (London: 1998), p. 135.
38. Cheryl Bentsen and Tom Field, 'Good Stuff Cheap' CIO: *The Magazine for Information Executives*, 1 December 2000, p. 174.
39. 'In December 1999 India world was bought by Satyam for 150 million dollars. This was one of the biggest dotcom acquisitions in India. That set everything through the roof.' Personal interview with Madanmohan Rao.
40. Personal interview with Ranjan Acharya and Anurag Behar (Bangalore, 20 January 2003).
41. Alam Srinivas Manish Khanduri, 'Software Hard Knocks', *Businessworld*, 26 March 2001.
42. Chandler, 'Asia's Businessmen of the Year: They Get IT'.
43. Personal interview with Ranjan Acharya and Anurag Behar.

6 Peripheral competencies

1. Recent alarmist reports in the US tend to conflate outsourcing with offshore outsourcing. The fact is that the majority of outsourcing contracts still go to US companies. Daniel Drezner and Glenn Reynolds quote fellow blogger Chuck Simmons: 'Most outsourced jobs don't go to India. They stay right here in the good, old U.S.A. That clerk from Accountemps or secretary from Kelly. That RN at your hospital. The cleaning crew in your office. Outsourced jobs.' (http://blog.simmins.org/2004_02_01_arch.html#107573066253829393).
2. Thomas Friedman, 'Tom's Journal', *The Online News Hour* (cited 10 March 2004); available from http://www.pbs.org/newshour/newshour_index.html.
3. Personal interview with Madanmohan Rao (Bangalore, 18 January 2003).
4. Raj Chengappa and Malini Goyal, 'Housekeepers to the World', *India Today*, 18 November 2002.
5. 'Stolen Jobs?', *The Economist*, 13 December 2003.
6. Ibid.
7. 'Outsourcing to India: Back Office to the World', *The Economist*, 5 May 2001.
8. Raj Chengappa, 'Housekeepers to the World'.
9. Shelley Singh and Mitu Jayashankar, 'The BPO Boom', *Businessworld*, 14 January 2002.

10. Pramod Bhasin (President GE Capital), Paper presented at the Hyderabad IT Forum, Hyderabad (2003).
11. 'America's Pain, India's Gain', *The Economist*, 11 January 2003. Wipro acquired Spectramind in order to enter the BPO space. Nipuna is the BPO subsidiary of Satyam.
12. Ritu Sarin, 'Software Superpower: India Banks on the Knowledge Trade', *Asia Week*, 7 April 2000.
13. Raj Chengappa, 'Housekeepers to the World'.
14. Personal interview with Animesh Thakur (Mumbai, 4 February 2003).
15. Raj Chengappa, 'Housekeepers to the World'.
16. 'Back Office to the World', *The Economist*, 3 May 2001.
17. Poorly managed companies have attrition rates ranging from 30 to 70 percent.
18. Personal interview with Animesh Thakur.
19. Personal interview with Brian Cravalaho (Bangalore, 21 January 2003).
20. Personal interview with Animesh Thakur.
21. Kripalani, 'The Rise of India'.
22. Managing this relationship, explains Ravindra Walters, can involve something as simple as explaining what 'yes' means. In India when someone says 'yes' to something the client might say it does not necessarily signify agreement. Yes might mean 'yes I have heard' not 'yes I have understood' or 'yes I will do it'. Personal interview with Ravindra Walters (Bangalore, 17 January 2003).
23. Personal interview with Animesh Thakur.
24. Ashish Gupta (Business Head, Hero MindMine), Paper presented at the Hyderabad IT Forum, Hyderabad (22–24 January 2003).
25. Personal interview with Animesh Thakur.
26. Personal interview with Brian Cravalaho.
27. Customers want to reduce the number of vendors they have to deal with and are seeking to bundle their contracts in software, IT services and BPO. The larger Indian vendors have recognized this and have moved into the BPO space. Personal interview with Ranjan Acharya and Anurag Behar (Bangalore, 2003).
28. Personal interview with Sunil Mehta (New Delhi, 13 January 2003).
29. The industry does, however, face some serious challenges. India is competing with the Philippines, China, Eastern Europe, Korea, Singapore, Ireland and Israel and though it is currently ahead of the pack it does have a number of potentially dangerous disadvantages. These include high telecom rates, long set-up times, an uncertain regulatory environment, convoluted licenses, problems achieving seamless connectivity and the endemic problem of India's terrible infrastructure.
30. Shelley Singh, 'The BPO Boom'.
31. Ibid.
32. 'Relocating the Back Office', *The Economist*, 11 December 2003.
33. Vidya Viswanathan, 'The Smiles Are Back', *Businessworld*, 9 December 2002.
34. Shelley Singh, 'The BPO Boom'.
35. Raj Chengappa, 'Housekeepers to the World'.
36. Daniel H. Pink, 'The New Face of the Silicon Age: How India Became the Capital of the Computing Revolution', *Wired*, 2004.
37. The cost of operations in India is currently 37 percent lower than in China and 17 percent lower than in Malaysia.
38. 'Stolen Jobs?'

39. Personal interview with Animesh Thakur. Local firms are working hard to foresee problems and develop solutions well in advance. For example, in order to mitigate against political and security risks, a war between India and Pakistan or a terrorist attack, local companies are setting up multiple locations creating second or third offices as risk aversion. This need to have a contingency plan has increased substantially after 9/11 and the Bali bombing when all companies want to hedge their bets. For this reason 'mobility across locations' is now seen as key to the continued success of the Indian BPO sector. BPO players in India are, thus, setting up contingency outfits offshore in the Philippines, China, US, Canada, Singapore and Malaysia. This has created a new kind of multinational company – which C.K. Prahalad has called Micro MNCs – that operates out of India but is global from the start.

40. Susan Sontag, 'The World as India', *Times Literary Supplement*, 6 June 2003.

41. Personal interview with Brian Cravalaho.

42. Call center employees working for an insurance company, for example, may have to deal with a customer who has recently lost his house, belongings or even family.

43. Personal interview with Brian Cravalaho.

44. Sontag, 'The World as India'.

45. Raj Chengappa, 'Housekeepers to the World'.

46. According to Cravalaho, this practice is shifting as people became accustomed to the idea that their call is being answered abroad. In places like England where there is a large South East Asian population, it was never really necessary in the first place. At any rate, for Cravalaho, these cultural concerns are dwarfed by the huge opportunity that the industry is creating. Personal interview with Brian Cravalaho.

47. Ibid.

48. Ibid.

49. Personal interview with Madanmohan Rao (Bangalore, 18 January 2003).

50. Ibid.

51. 'Back Office to the World'.

52. Tom Field 'For a Few Rupees More' in Cheryl Bentsen and Tom Field, 'India Unbound: CIO Field Report', *CIO: The Magazine for Information Executives*, 1 December 2000, p. 168.

53. Ibid.

54. Sunil Mehta of Nasscom questions the relevance of the notion of a value chain. 'In this industry there is no concept of value chain', he explains. 'A value chain means that the higher I move the more money I make. Yet Accenture who are doing the highest work imaginable have a profit margin of 3 per cent, at Infosys the profit margin is 29 per cent. So who is where on the value chain? Indians are very good at so called "low end work." So what? Somebody has got to do it.' Personal interview with Sunil Mehta.

55. Many advocates of free trade have singled out Lou Dobbs' protectionist rhetoric as particularly noxious. *The Economist* magazine in a 'leader' on outsourcing writes that Dobbs 'greets every announcement of lost jobs as akin to a terrorist assault'. 'The New Jobs Migration', *The Economist*, 21 February 2004. While the online journal *Tech Central Station* has developed what amounts to an anti-Dobbs campaign. The most amusing – and devastating – aspect of this is the 'Lou Dobbs Rogue Fund' which draws

on a list of companies compiled by Dobbs, which he singles out in order to reprimand their outsourcing practice. James K. Glassman, host of *Tech Central Station*, suggests that the list makes an excellent stock portfolio and bets that it will 'beat the market as a whole'. James K. Glassman, *The Dobbs Rogue Fund* (Tech Central Station, 2004 [cited]); available from http://www.techcentralstation.com/021704C.html.

56. Viswanathan, 'The Smiles are Back'.
57. 'Stolen Jobs?'
58. Pink, 'The New Face of the Silicon Age: How India Became the Capital of the Computing Revolution'.
59. Kripalani, 'The Rise of India'.
60. Vivek Agrawal and Diana Farrell, 'Who Wins in Offshoring', *McKinsey Quarterly*, 2003, p. 2.
61. The blogosphere has done a good job of stating the merits of offshore outsourcing. A collection of articles can be accessed through Daniel Drezner's blog at http://www.danieldrezner.com/blog/.
62. Personal interview with Sunil Mehta.
63. 'Stolen Jobs?'
64. Agrawal, 'Who Wins in Offshoring'. No doubt there is painful job loss from outsourcing. McKinsey reports that though 69 percent found new work this still leaves a huge number who did not. Moreover of those who do find new jobs 55 percent took a pay cut, while 25 percent took paycuts of 30 percent or more. The McKinsey report suggests that some of the gains from free trade help these un- or under-employed.
65. 'Even at their peak in 2001, the number of "trade-related" layoffs represented a mere 0.6 per cent of American unemployment.' 'The Great Hollowing-out Myth', *The Economist*, 2004.
66. Kripalani, 'The Rise of India'.
67. Manoj Chandran, *India Not to Be Perturbed by New Jersey Bill* [website] (CIOL IT Unlimited, 2002 [cited 2 February 2004]); available from http://www.ciol.com/content/news/trends/102123001.asp.
68. Editorial, *Financial Times*, 21 February 2003.
69. 'While it's easy to see why labor unions might oppose this sort of thing [the outsourcing of jobs], it's hard for me to see it as a liberal issue, really. After all, aren't liberals supposed to be *for* the redistribution of wealth from the better-off to the less-well-off? These jobs don't disappear, after all: they go overseas, to people who probably need them more. Isn't that a *good thing*? Or, at least, to me it's not obviously worse than, say, taxing corporations in a way that causes them to cut jobs, and then using the money to pay for foreign aid. In fact, it's probably better, overall, since it builds up a corps of educated professionals in other countries, instead of fostering the sort of dependency (and corruption) that usually results from foreign aid' Glenn Reynolds, *Outsourcing Elections* (Tech Central Station, 2003); available from http://www.techcentralstation.com/062503A.html.
70. Pink, 'The New Face of the Silicon Age: How India Became the Capital of the Computing Revolution'.
71. Editorial.
72. Video clip of Bill Clinton defending NAFTA reproduced by William Cran and Daniel Yergin, 'The Agony of Reform', in *Commanding Heights: The Battle for the World Economy* (WGBH, 2003).

73. Rajeev Dubey, 'India as a Global R&D Hub', *Businessworld*, 17 February 2003.
74. P. Hari, 'Bangalore Technopolis', *Businessworld*, 26 February 2001.
75. Dubey, 'India as a Global R&D Hub'.
76. Clay Chandler, 'Asia's Businessmen of the Year: They Get IT', *Fortune*, 17 February 2003.
77. Shelley Singh, 'BPO. Cover Story', *Businessworld*, 19 August 2002.
78. Don Tapscott, Paper presented at the Nasscom India Leadership Forum, Mumbai (11–14 February 2003).
79. Singh, 'BPO. Cover Story'.
80. Tapscott.
81. Singh, 'BPO. Cover Story'.
82. Tapscott.
83. Kripalani, 'The Rise of India'.

7 The digital dividend

1. Statistics obtained from *How Many Online?* (Nua: The World's leading resource for Internet trends and statistics, 2002 [cited 2 February 2004]); available from http://www.nua.ie/surveys/how_many_online/index.html.
2. Celia Dugger, 'Connecting Rural India into the World', *New York Times*, 28 May 2000.
3. Aruna Sundarajan, Paper presented at the Hyderabad IT Forum, Hyderabad (22–24 January 2003).
4. Allan Hammond, 'Digitally Empowered Development', *Foreign Affairs* (2001).
5. *The Economist* cites a new paper by Carsten Fink and Charles Kenny, two economists at the World Bank, which argues that while the digital divide between the developed and developing world appears to be shrinking, there is increasing concern about the digital divide within developing countries like India. 'Canyon or Mirage?', *The Economist*, 24 January 2004.
6. C.K. Prahalad, *India as a Source of Innovations: The First Lalbahadur Shastri National Award for Excellence in Public Administration and Management Sciences Lecture*, September 30, 2000, New Delhi (Digital Dividend, 2000 [cited 2 February 2004]); available from www.digitaldividend.org/pdf/0203ar03.pdf.
7. Narayana Murthy, Paper presented at the Nasscom, India Leadership Forum, Mumbai (11–14 February 2003).
8. Bhoomi's website can be accessed at http://www.revdept-01.kar.nic.in/.
9. 'Mirage, or Reality?', *20 Years: 1982–2002. Dataquest: Collectors Issue*, 15 December 2002.
10. According to Rajeev Chawla, if e-government schemes make fighting corruption an explicit goal they will be met with an inordinate amount of resistance. As an example, he cited the fact that Bhoomi was using simputers for crop data collection. Fearing the transparency IT can produce, the village accountants did everything in their power to ensure they would not work. Rajeev Chawla, Paper presented at the Hyderabad IT Forum (22–24 January 2003).
11. This has the surplus value of proving that it is possible for IT innovation to move in unexpected directions, from the poor rural regions to the global high-tech cities.

12. ITC's website describes this vicious cycle as follows: 'low risk taking ability > low investment > low productivity > weak market orientation > low value addition > low margin > low risk taking ability'. http://www.itcibd.com/e-choupal1.asp.
13. e-Choupal's website can be accessed at http://www.echoupal.com/default.asp.
14. Data is available through ITC's e-Choupal website available at http://www.itcibd.com/e-choupal1.asp.
15. Quoted from e-Choupal's website.
16. Anupama Katakam, 'The Warana Experiment', *Frontline*, 5–18 January 2002.
17. Ibid.
18. Madanmohan Rao (ed.), 'I-India: The Hope Amidst the Hype', in *The Asia-Pacific Internet Handbook. Episode IV: Emerging Powerhouses* (New Delhi: Tata McGraw-Hill, 2002), p. 224.
19. Gurcharan Das, *India Unbound* (New Delhi; New York: Viking, 2000), p. 208.
20. Ibid., p. 209.
21. Rao, 'I-India: The Hope Amidst the Hype', p. 204.
22. Sam Pitroda, Paper presented at the TiEcon, New Delhi (8 January 2003).
23. Prasanto K. Roy, 'Foundation and Empire', *20 Years: 1982–2002. Dataquest Collectors Issue*, 2002.
24. Dugger, 'Connecting Rural India into the World'.
25. 'Getting India to Go Online', *Businessworld*, 31 January 2000.
26. J. Satyanarayana (IT secretary Andhra Pradesh), Paper presented at the Hyderabad IT Forum, Hyderabad (22–24 January 2003).
27. The main reason for this is an unfriendly business environment, extremely strong unions and a resultant lack of entrepreneurialism. Billboards posted around the state urging people to make the business climate as warm as the weather is evidence of the fact that the government is at least aware of the problem.
28. ICT is able to create access to otherwise isolated communities, open up distribution and market opportunities as well as providing people with secure identities, credit history and even a virtual address, all of which are needed in order to participate in the world economy.
29. Rakesh Khanna, 'Tarahaat: Achieving Connectivity for the Poor: Case Study', *Digital Partners*, [cited 3 February 2004]; available from http://www.digitalpartners.org/tara.html.
30. From SARI's website available at http://edev.media.mit.edu/SARI/mainsari.html.
31. http://gyandoot.nic.in/.
32. The story of Gyandoot is told in Rajesh Rajora, *Bridging the Digital Divide: Gyandoot the Model for Community Networks* (New Delhi: Tata McGraw-Hill, 2002).
33. Information obtained from the products section of Telecommunication and Computer Networks Group (TeNeT) available at http://www.tenet.res.in/Products/tele_prod.html.
34. http://www.digitaldividend.org/.
35. Personal interview with Madanmohan Rao (Bangalore, 18 January 2003).
36. Prahalad, *India as a Source of Innovations: The First Lalbahadur Shastri National Award for Excellence in Public Administration and Management Sciences Lecture*, 30 September 2000, New Delhi.

37. 'According to my good friend, Prof. C.K. Prahalad of Core Competence fame, the Third World is just a state of mind rather than any lack of resource. I could not agree more with him.' N.R. Narayana Murthy, 'It is all in the Mind, Stupid', *Business Today*, 19 January 2003.

38. Prahalad, *India as a Source of Innovations: The First Lalbahadur Shastri National Award for Excellence in Public Administration and Management Sciences Lecture*, 30 September 2000, New Delhi.

39. Ibid.

40. C.K. Prahalad and Stuart Hart, 'The Fortune at the Bottom of the Pyramid', *Digital Dividend*, 2002; available from http://www.digitaldividend.org/pubs/pubs.htm.

41. C.K. Prahalad and Allen Hammond, *What Works: Serving the Poor, Profitably* (Digital Dividend, 2002 [cited 2 February 2004]); available from http://www.digitaldividend.org/pubs/pubs.htm.

42. Ibid.

43. Hernando de Soto, *The Mystery of Capital: Why Capitalism Triumphs in the West and Fails Everywhere Else* (New York: Basic Books, 2000), p. 35.

44. Ibid., p. 37.

45. Prahalad, *What Works: Serving the Poor, Profitably*.

46. The poor pay huge amounts for credit. It is common to pay 40–70 per cent interest rates to even non-profit lending institutions.

47. Microloans and microcredit are very expensive due to the transaction costs. Citigroup is experimenting with developing banking kiosks designed to serve small depositors.

48. Prahalad, *What Works: Serving the Poor, Profitably*.

49. C.K. Prahalad, Paper presented at the TiEcon, New Delhi (8 January 2003).

50. Hammond, 'Digitally Empowered Development'.

51. Ibid.

52. Ibid.

53. Prahalad.

54. Hart, *The Fortune at the Bottom of the Pyramid*. 'The opposing perception that they do not constitute a market,' Hart and Prahalad continue, 'ignores the growing importance of the informal economy that accounts for 40–60 per cent of all economic activity in all developing countries.'

55. Ibid.

56. Hart, *The Fortune at the Bottom of the Pyramid*.

57. Hernando de Soto interviewed in Yergin, 'The New Rules of the Game'.

58. Prahalad, *What Works: Serving the Poor, Profitably*.

59. Ibid.

60. Ibid.

61. Prahalad, *What Works: Serving the Poor, Profitably*.

62. Hart, *The Fortune at the Bottom of the Pyramid*.

63. Prahalad.

64. Prahalad, *What Works: Serving the Poor, Profitably*.

65. Prahalad, *India as a Source of Innovations: The First Lalbahadur Shastri National Award for Excellence in Public Administration and Management Sciences Lecture*, 30 September 2000, New Delhi.

66. Partha Iyengar, Paper presented at the Hyderabad IT Forum (22–24 January 2003).

67. Ramalinga Raju, Paper presented at the Nasscom India Leadership Forum, Mumbai (11–14 February 2003).
68. Clayton Christenson, Thomas Craig and Stuart Hart, 'The Great Disruption', *Foreign Affairs* (2001), p. 81.
69. Joseph Alois Schumpeter, *Capitalism, Socialism, and Democracy*, 2nd edn (New York; London: Harper & Brothers, 1947), p. 82.
70. Ibid., p. 83.
71. Friedman quoting James Surowiecki in Thomas L. Friedman, *The Lexus and the Olive Tree* (New York: Anchor Books, 2000), p. 11.
72. Schumpeter, *Capitalism, Socialism, and Democracy*, p. 83.
73. For more on this theme, see Clayton Christenson, *The Innovator's Dilemma* (HarperCollins, 2000).
74. Christenson, 'The Great Disruption', p. 81.
75. Mukesh Ambani, 'Preparing the Indian Mind for the New World', *Business Today*, 19 January 2003.

8   Global networks

1. 'We went through an elaborate election process', says Kanwal Rekhi, 'all the twenty chapters participated. We had a constitutional convention and picked Professor Prahalad to define the process. That took about a year and then we spent another year drafting the initial constitution and election process and that process finished last October.' Personal interview with Kanwal Rekhi (Silicon Valley, 17 June 2002).
2. Personal interview with Kailash Joshi.
3. From TiE's website available at http://www.tie.org/.
4. Personal interview with Kanwal Rekhi.
5. Ibid. 'More recently TiE is explicitly aiming to reach out to the mainstream and other ethnic groups and has staged events with Israeli and Chinese engineers in the Valley. Due to this openness TiE', claims Rekhi, 'is now seen as a group worthy of emulation by other ethnic groups.'
6. Personal interview with Kailash Joshi.
7. From TiE's promotional material passed out at TiEcon 2002.
8. See the 'About TiE' section of the organization's website available at http://www.tie.org/site/About/About%20TiE.
9. From 'TiE News and Views' promotional material passed out at TiEcon 2002.
10. Personal interview with Professor Anand Patwardhan (Mumbai, 6 February 2003).
11. Gurcharan Das, *India Unbound* (New Delhi; New York: Viking, 2000), p. 261.
12. 'Ramp-up Mode', *Business Today*, 23 June 2002.
13. Alam Srinivas, 'A Debugged Operating $ystem', *Outlook*, 27 January 2003.
14. Clay Chandler, 'Asia's Businessmen of the Year: They Get IT', *Fortune*, 17 February 2003.
15. Ibid.
16. Ibid.
17. AnnaLee Saxenian, *Silicon Valley's New Immigrant Experience* (California: Public Policy Institute of California, 1999), p. 49.
18. Kanwal Rekhi quoted in *Upside Today*, 28 November 2000.

19. Kanwal Rekhi quoted in Gurmeet Naroola, *The Entrepreneurial Connection: East Meets West in Silicon Valley* (California USA: Special TiE Edition, 2001), p. 93.
20. I. Copeland, 'A Capital Idea for the High Tech Elite', *Washington Post*, Friday, 26 May 2000.
21. Ibid.
22. Anthony Spaeth, 'The Golden Diaspora', *Time*, 19 June 2000.
23. Ibid.
24. Mira Kamdar, 'Reinvinting India', *the-south-asian.com*, 2001.
25. Melanie Warner, 'The Indians of Silicon Valley', *Fortune Magazine*, 15 May 2000.
26. Sue Birley, 'The Role of Networks in the Entrepreneurial Process', *Journal of Business Venturing* 1 (1985), p. 109.
27. Antoine Pecoud, 'Thinking and Rethinking Ethnic Economies', *Diaspora* 9: 3 (2000), p. 454.
28. Ibid., p. 458.
29. Ibid., p. 456.
30. Ibid., p. 455.
31. The theorist Manuel Delanda draws on Saxenian's work to argue that the differences between the two regions correspond to the distinction made by historian Fernand Braudel between markets and capitalism or – as Delanda prefers – markets and anti-markets. Route 128 is an 'antimarket' dominated by a small number of large corporations that operate through a centralized command and have intimate links to the state. Silicon Valley – the region at the heart of the information economy – has much more in common with markets in this true Braudelian sense of the term. The Valley is a 'noisy heterogenous system', claims Delanda, in which a collection of different sized enterprise have formed an interactive network of small producers which function with a decentralized decision-making process. Manuel Delanda, 'Markets, Antimarkets and Network Economics', paper presented at the Virtual Futures (Warwick University, UK, 1996). As Rekhi says 'the economy is reverting to what is natural dominated by individualistic cobblers, carpenters and ironsmiths rather than globocorporations like IBM and General Motors'. Vidya Viswanathan, 'Indian Internet Mafia', *Businessworld*, 24 May 1999.
32. See AnnaLee Saxenian, *Regional Advantage: Culture and Competition in Silicon Valley and Route 128* (Boston: Harvard University Press, 1996).
33. TiE promotional material handed out at TiEcon 2002.
34. Saxenian, *Regional Advantage: Culture and Competition in Silicon Valley and Route 128*, p. 33.
35. Ibid.
36. Constance Lever-Tracy, David Fu-Keung Ip and Noel Tracy, *The Chinese Diaspora and Mainland China: An Emerging Economic Synergy* (Houndmills; Basingstoke; Hampshire; New York: Macmillan Press; St Martins Press, 1996).
37. Joel Kotkin, *Tribes: How Race, Religion, and Identity Determine Success in the New Global Economy*, 1st edn (New York: Random House, 1993), p. 17.
38. Ibid., p. 89.
39. Ibid.
40. Ibid., p. 70.
41. James C. Bennett, *An Anglosphere Primer*. Presented to the Foreign Policy Research Institute, ©2001 (2002); available from http://www.pattern.com/bennettj-anglosphereprimer.html.

42. Ibid.
43. Ibid.
44. Ibid.
45. Ibid.
46. Ibid.
47. Ibid.
48. Braj B. Kachru, *The Other Tongue: English across Cultures* (Urbana: University of Illinois Press, 1982), p. 8.
49. Samuel P. Huntington, *The Clash of Civilizations and the Remaking of World Order*, 1st Touchstone edn (New York: Touchstone, 1997), p. 62.
50. Braj B. Kachru, *The Indianization of English: The English Language in India* (Delhi; New York: Oxford, 1983), p. 127.
51. Huntington, *The Clash of Civilizations and the Remaking of World Order*, p. 61.
52. Ibid.
53. Ibid., p. 62.
54. In order to grasp the power of this open strategy it is useful to compare the story of English with the case of French, 'a language that had more speakers than English as recently as the eighteenth century'. Kotkin, *Tribes: How Race, Religion, and Identity Determine Success in the New Global Economy*, p. 75.

    The French have long pursued a policy of strict cultural protectionism. Since as early as 1635, French has been rigorously controlled by the Academie Francaise in Paris, an organization dedicated to preserving the language and fixing its rules. As Joel Kotkin writes, 'the "universality" of French language and culture implied one shared set of norms which, of course, could be regulated – and preserved from adulteration – by the Academie Francaise'. Ibid.

    The authoritarian rule of the Academie is designed to protect the French language from any alien influences. In today's world the most threatening of these is of course English. For this reason 'in official contexts it is illegal to use an English word where a French word exists, even though the usage may have widespread popular support (e.g. computer for ordinateur)'. David Crystal, *English as a Global Language* (Cambridge, England; New York: Cambridge University Press, 1997).
55. Those standards that do exist generally come from the media, for example RP (or received pronunciation) which is associated with the BBC, or as an appeal for linguistic tolerance (i.e. GA – General American, which is not so much a model as a demand for the acceptance of the deviations of American-English). The Oxford English Dictionary – a type of standard – is itself having to catch up with bottom-up innovation. Kachru points out that 'one interesting aspect of Standard English is that in every English-using community those who habitually use only standard English are in a minority'. Kachru, *The Other Tongue: English across Cultures*, p. 34.
56. Robert MacNeil, Robert McCrum and Wiliam Crun, *The Story of English* (London: Faber & Faber and BBC Books, 1992), pp. 43–44.
57. Huntington, *The Clash of Civilizations and the Remaking of World Order*, p. 62.
58. Peter Strevens, 'The Localized Forms of English', in *The Other Tongue: English across Cultures*, ed. Braj B. Kachru (Urbana: University of Illinois Press, 1982), p. 24.
59. Ibid.

60. Kachru, *The Other Tongue: English across Cultures*, p. viii.
61. Madelaine Drohan and Alan Freeman, 'English Rules', in *Globalization and the Challenges of a New Century: A Reader*, ed. Howard D. Mehlinger and Patrick O'Meara, Mathew Krain (Bloomington: Indiana University Press, 2000), p. 428.
62. Kachru, *The Other Tongue: English across Cultures*, p. 2.
63. Ibid.
64. Braj B. Kachru, *The Alchemy of English: The Spread, Functions, and Models of Non-Native Englishes*, 1st edn, *English in the International Context* (Oxford [Oxfordshire]; New York: Pergamon Institute of English, 1986), p. 20.
65. 'English is not part of the government's Central Institution of Indian Languages (because it is not an Indian Language) nor does it belong to the Central Institute of "Other Foreign Languages" (because it is not a foreign language).' Archana S. and N. Krishnaswamy Burde, *The Politics of Indians' English* (Oxford: Oxford University Press, 1998), p. 44.
66. Anthony Burgess quoted in MacNeil, *The Story of English*, p. 355.
67. Kachru, *The Indianization of English: The English Language in India*, p. 2.
68. For the rules governing Indian English, see R.S. Gupta and Kapil Kapoor, eds, *English in India Issues and Problems* (Academic Foundation, 1991), pp. 197–198, for the structure and rules governing the code-mixing of Hinglish see Gupta, ed., *English in India Issues and Problems*, p. 210.
69. Kachru writes of an IIT professor Dr M.P. Jain who – with the aid of a computer – is studying the unique character of Indian English. Dr Jain emphasizes four important characteristics of Indian English: (1) the use of archaic words (i.e. timepiece); (2) using words borrowed from Indian words (e.g. lakhs); (3) combining two English words (e.g. eve-teaser); and (4) Indian-English hybrids (e.g. newspaper wallah). He also notes that the Indianization of English is not restricted to words and phrases but also includes grammatical constructions.
70. R.S. Gupta, 'English in Post-Colonial India', in *Who's Centric Now? The Present State of Post-Colonial Englishes*, ed. Bruce Moore (Oxford: Oxford University Press, 2001), p. 155.
71. Gupta, ed., *English in India: Issues and Problems*, p. 208.
72. Ibid.
73. Quoted in Mehrotra Raja Ram, *Indian English Texts and Interpretations* (Amsterdam, Philadelphia: John Benjamins Publishing Company, 1998), p. 6.
74. Kachru, *The Alchemy of English: The Spread, Functions, and Models of Non-Native Englishes*, p. 12.
75. Salman Rushdie quoted in MacNeil, *The Story of English*, p. 34.
76. There are several reasons for this, including problems with a wireless network standard in North America, cheap land lines and a general lack of mobile phones.
77. The only English-speaking region in the West with widespread use of SMS is the UK.
78. Shefalee Vasudev, 'Love, Sex and SMS', *India Today*, 14 October 2002.

9 Zero logo

1. Rituraj Nath, a Nasscom executive, made it clear in an interview in 1999 that branding India was one of the organization's primary goals.
2. Personal interview with Sunil Mehta (New Delhi, 13 January 2003).

3. Bruce Einhorn, India and IT: 'Like France and Wine' (*Advantage India*, 28 January 2002 [cited 5 February 2004]); available from http://www. nasscom.org/artdisplay.asp?Art_id = 1415.
4. Ibid.
5. The Cool Britannia campaign was criticized at home for its top-down approach and excessively close ties to the 'New Labour' party of Tony Blair.
6. For more on national branding, see Wally Olins, *Trading Identities: Why Countries and Companies Are Taking on Each Others' Roles* (London: The Foreign Policy Centre, 1999).
7. Personal interview with Kailash Joshi (Silicon Valley, 17 June 2002).
8. Nasscom, 'The Software Industry in India: A Strategic Review', p. 34.
9. Gurcharan Das, *India Unbound* (New Delhi; New York: Viking, 2000), p. 174.
10. 'Man with a Mission: Interview with D Mehta', *Financial Times*, 1999.
11. The word Juggernaut comes from the Indian tribal god 'Lord Jagannath' who is primarily worshipped in Puri, Orissa. Every year, giant icons of him and his consort are placed on huge chariots and taken out in procession. The Franciscan missionary, Friar Odoric, who visited India in the 14th century, wrote in his journal about the Jagannath's ritual procession and how the devotees threw themselves at the chariot wheels allowing themselves to be crushed. According to Odoric, the people were offering themselves as sacrifice to the Lord. Many Indians deny this, saying that any deaths were accidental, caused by the enormous crowds of worshippers.
12. In a somewhat ironic twist, India has hired 'Hill and Knowlton', a leading PR company to run a marketing strategy in order to counter the backlash against outsourcing.
13. Tulasi Srivanas, 'A Tryst with Destiny: The Indian Case of Cultural Globalization', in *Many Globalizations: Cultural Diversity in the Contemporary World*, eds Samuel P. Huntington and Peter L. Berger (New York: Oxford University Press, 2003), p. 90.
14. John McLeish, *Number*, 1st American edn (New York: Fawcett Columbine, 1992), p. 115.
15. Georges Ifrah, *The Universal History of Numbers: From Prehistory to the Invention of the Computer*, trans. E.F. Harding David Bellos, Sophie Wood and Ian Monk (New York: John Wiley & Sons, Inc., 2000), p. 346.
16. Ibid., p. 510.
17. Ibid., p. 245.

'It is impossible to exaggerate the essential importance of zero,' writes Ifrah, which did not simply allow the empty positions in the place value system to be expressed; nor did it just provide a word, a figure or a symbol. Above all, it created a notion, understood at once as a numerical concept and an arithmetical operator, and as a real number inversely equal to mathematical infinity and simultaneously a member of the sets of integers of fractions, of real numbers, of complex numbers and so on, from one generalization to the next ... This fundamental concept is positioned at the meeting point of all the branches of modern mathematics: arithmetic of integers and fractions, the algebras of scalar and vector quantities, tensor calculus, matrix calculus numerical analysis, infinitesimal analysis, differential and integral calculus, set theory, algebraic and topological structures, etc. And it naturally plays just such a fundamental role in all the other scientific disciplines, from physics, astronomy, astrophysics, statistics, economics or econometrics to biology, chemistry, linguistics, computing robotics and cybernetics.

18. Ibid., p. 345.
19. Ibid., p. 594.
20. Sumero-Babylonian sexigesimal numeracy supported a chronogeometric system, in which an Ideal 360 day model of the year was mapped onto the zodiacal circle. Modern clock-time and geometry still count in this way, subdividing hours and geometric degrees into minutes and seconds.
21. Charles Seife, *Zero: The Biography of a Dangerous Idea* (New York: Viking, 2000), p. 15.
22. Ifrah, *The Universal History of Numbers: From Prehistory to the Invention of the Computer*, p. 342.
23. The Mayans also had a place value system and so a symbol for zero. Mayan civilization (500–925 CE) reached amazing levels of arithmetical sophistication, especially in their timekeeping practices. The highly ritualistic Mayan calendar with its complex cycles that marked the intricate rhythm of Mayan civilization also incorporated a zero sign.
24. Ifrah, *The Universal History of Numbers: From Prehistory to the Invention of the Computer*, p. 346.
25. McLeish, *Number*, pp. 115–116.
26. John D. Barrow, *The Book of Nothing* (London: Jonathan Cape, 2000), p. 37.
27. Ibid., p. 37.
28. Ibid., p. 38.
29. Seife, *Zero: The Biography of a Dangerous Idea*, p. 2. These bizarre, almost incomprehensible results, lead Seife to claim that by 'dividing by zero – just once – . . . destroys the entire framework of mathematics'. Ibid., p. 23.
30. Florian Cajori, *A History of Mathematics* (New York: Macmillan, 1894), p. 100.
31. Robert Logan quoted in Barrow, *The Book of Nothing*, p. 13.
32. Seife, *Zero: The Biography of a Dangerous Idea*, p. 34.
33. Gilles Deleuze and Félix Guattari, *A Thousand Plateaus*, trans. Brian Massumi (Minneapolis: University of Minnesota Press, 1987), p. 389.
34. Proclus and Glenn R. Morrow, *A Commentary on the First Book of Euclid's Elements* (Princeton, NJ: Princeton University Press, 1970), p. 52.
35. Deleuze and Guattari, *A Thousand Plateaus*, p. 389.
36. Ibid., p. 388.
37. Cajori, *A History of Mathematics*, p. 85.
38. Seife, *Zero: The Biography of a Dangerous Idea*, p. 70.
39. Ibid.
40. Barrow, *The Book of Nothing*, p. 42.
41. In the shunya there is no form, no sensation, there are no ideas, no volitions and no consciousness. In the shunya there are no eyes, no ears, no nose, no tongue, no body, no mind. In the shunya, there is no color, no smell, no taste, no contact and no elements. In the shunya there is no ignorance, no knowledge or even the end of ignorance. In the shunya there is no aging or death. In the shunya there is no knowledge, or even the acquisition of knowledge. Ibid., p. 496.
42. Ifrah, *The Universal History of Numbers: From Prehistory to the Invention of the Computer*, p. 507.
43. Barrow, *The Book of Nothing*, p. 42.

44. Seife, *Zero: The Biography of a Dangerous Idea*, p. 65. This cyclical frame of mind is found in the Indian philosophy of time, or Yuga doctrine. The Yugas, four in number, decimally divide a Mahayuga (or 'Great Yuga', of 4,320,000 years) in the ratio 4:3:2:1. The accumulation of Yugas involves both cyclic repetition and serial rupture, registered by strings of terminal zeroes, each ciphering the simultaneous completion of gradual process and sudden transition to a new epoch.
45. Ifrah, *The Universal History of Numbers: From Prehistory to the Invention of the Computer*, p. 495.
46. Georges Ifrah, *The Universal History of Computing: From the Abacus to the Quantum Computer* (New York: John Wiley, 2001), pp. 244–245.
47. Ibid.
48. Barrow, *The Book of Nothing*, p. 44.
49. Ibid., p. 40.
50. Ifrah, *The Universal History of Numbers: From Prehistory to the Invention of the Computer*, p. 510.
51. Seife, *Zero: The Biography of a Dangerous Idea*, p. 2.
52. Ibid., p. 63.
53. Ibid., p. 39.
54. Plato quoted in Barrow, *The Book of Nothing*, p. 45.
55. Seife, *Zero: The Biography of a Dangerous Idea*, p. 25.
56. Ibid.
57. John Milton, *Paradise Lost and Paradise Regained*, The Signet Classic Poetry Series (New York: New American Library Inc, 1968), pp. 93–94.
58. Barrow, *The Book of Nothing*, p. 42.
59. Ibid.
60. Ifrah, *The Universal History of Numbers: From Prehistory to the Invention of the Computer*, p. 588.
61. Barrow, *The Book of Nothing*, p. 74.
62. Ibid.
63. Seife, *Zero: The Biography of a Dangerous Idea*, p. 23.
64. Ifrah, *The Universal History of Numbers: From Prehistory to the Invention of the Computer*, p. 588.
65. Ibid., p. 510.
66. The word Algorism comes from al-Khurwarizmi, the first Islamic scholar who generalized their application. 'It was through translations from Arabic that the West eventually became familiar with the works of Euclid, Archimedes, Ptolemy, Aristotle, al Khuwarizimi, al Biruni, Ibn Sina, and many others.' Ibid.
67. Ibid., p. 586.
68. Ibid., p. 588.
69. Barrow, *The Book of Nothing*, p. 49.
70. Ibid., p. 48.
71. Ifrah, *The Universal History of Numbers: From Prehistory to the Invention of the Computer*, p. 577.
72. Seife, *Zero: The Biography of a Dangerous Idea*, p. 25.
73. The Gregorian calendar was adopted by the protestant German states in 1699, England and its colonies in 1752, Sweden in 1873, Japan in 1873, China in 1912 and the Soviet Union in 1918.
74. McLeish, *Number*, p. 5.

75. Seife, *Zero: The Biography of a Dangerous Idea*, p. 2.
76. Conrad Joseph, *The Secret Agent* (Hertfordshire: Wordsworth Classics, 1993), p. 35.
77. Ibid., p. 36.
78. Barrow, *The Book of Nothing*, p. 46.
79. Ifrah, *The Universal History of Numbers: From Prehistory to the Invention of the Computer*, p. 594.
80. Benjamin Barber, 'Jihad Versus McWorld', in *Globalization and the Challenges of the New Century: A Reader*, ed. Howard D. Mehlinger, Mathew Krain and Patrick O'Meara (Bloomington Indiana University Press, 2000), p. 26.

Bibliography

Agrawal, Vivek and Diana Farrell, 'Who Wins in Offshoring', *McKinsey Quarterly 4* (2003).

Allen Hammond and C.K. Prahalad, *What Works: Serving the Poor, Profitably*, Digital Dividend (2002); available from http://www.digitaldividend.org/pubs/pubs.htm.

Ambani, Mukesh, 'Preparing the Indian Mind for the New World' in *Business Today*, 19 January 2003, p. 126.

'America's Pain, India's Gain', *The Economist*, 11 January 2003.

Amin, Samir, *Eurocentrism* (New York: Monthly Review Press, 1989).

Avineri, Shlomo, ed., *Karl Marx on Colonialism and Modernization: His Dispatches and Other Writings on China, India, Mexico, the Middle East and North Africa* (New York: Doubleday, 1968).

'Back Office to the World', *The Economist*, 3 May 2001.

Balasubramanyam, V.N., *Conversations with Indian Economists* (London: Palgrave, 2001).

Barabasi, Albert-Laszlo, *Linked: The New Science of Networks* (New York: Perseus, 2002).

Barber, Benjamin, *Jihad vs McWorld: How Globalism and Tribalism Are Re-Shaping the World* (New York: J. Ballantine Books, 1996).

——, 'Jihad Versus McWorld' in *Globalization and the Challenges of the New Century: A Reader*, edited by Howard D. Mehlinger, Mathew Krain and Patrick O'Meara (Bloomington: Indiana University Press, 2000).

Barrow, John D., *The Book of Nothing* (London: Jonathan Cape, 2000).

Barua, Apratim, 'Cyber-Coolies, Hindi and English; Letter', *Times Literary Supplement*, 1 August 2003.

Bear, Greg [*Slant*], 1st edn (New York: Tor, 1997).

Bennett, James C., *An Anglosphere Primer*, Presented to the Foreign Policy Research Institute, ©2001 (2002); available from http://www.pattern.com/bennettj-anglosphereprimer.html.

Bentsen, Cheryl and Tom Field, 'India Unbound: CIO Field Report', *CIO: The Magazine for Information Executives*, 1 December 2000, pp. 90–206.

Bhagwati, Jagdish, 'Why Your Job Isn't Moving to Bangalore', *New York Times*, 15 February 2004.

Bhasin, Pramod, Paper presented at the Hyderabad IT Forum, Hyderabad (22–24 January 2003).

Bhatt, Kamla, 'IITs', *Outlook*, 3 February 2003, pp. 42–43.

Birley, Sue, 'The Role of Networks in the Entrepreneurial Process', *Journal of Business Venturing* 1 (1985), pp. 107–117.

Bonacich, Edna, 'A Theory of Middleman Minorities', *American Sociological Review* 38 (1973), pp. 583–594.

'BPO Ladder', *Businessworld*, 14 January 2002, pp. 30–37.

Braudel, Fernand, *The Perspective of the World: Civilization and Capitalism 15th to 18th Century* (London: Phoenix Press, 1984).

'Budget 1999–2000 Highlights: An Analysis of Impact on Computer Software Industry' (New Delhi: Nasscom, 1999).

Burde, Archana S. and N. Krishnaswamy, *The Politics of Indians' English* (Oxford: Oxford University Press, 1998).

Cajori, Florian, *A History of Mathematics* (New York: Macmillan, 1894).

'Canyon or Mirage?' *The Economist*, 24 January 2004, p. 70.

Chandler, Clay, 'Asia's Businessmen of the Year: They Get IT', *Fortune*, 17 February 2003, pp. 38–42.

Chandran, Manoj, *India Not to Be Perturbed by New Jersey Bill* [website], CIOL IT Unlimited (2002); available from http://www.ciol.com/content/news/trends/102123001.asp.

Chawla, Rajeev, Paper presented at the Hyderabad IT Forum, Hyderabad (22–24 January 2003).

Chengappa, Raj and Malini Goyal, 'Housekeepers to the World', *India Today*, 18 November 2002, pp. 38–49.

'China Crises', *The Economist*, 7 June 2003.

Christenson, Clayton, *The Innovator's Dilemma* (New York: HarperCollins, 2000).

——, Thomas Craig and Stuart Hart, 'The Great Disruption', *Foreign Affairs* (March/April 2001), pp. 80–95.

Cohen, Robin, *Global Diasporas: An Introduction* (Seattle: University of Washington Press, 1997).

Conrad, Joseph, *The Secret Agent* (Hertfordshire: Wordsworth Classics, 1993).

Copeland, L., 'A Capital Idea for the High Tech Elite', *Washington Post*, Friday, 26 May 2000.

Crystal, David, *English as a Global Language* (Cambridge, England; New York: Cambridge University Press, 1997).

Das, Gurcharan, 'Cyber-Coolies, Hindi and English; Letter', *Times Literary Supplement*, 12 September 2003.

——, *India Unbound* (New Delhi; New York: Viking, 2000).

——, *The Elephant Paradigm: India Wrestles with Change* (New Delhi: Penguin, 2002).

Dasgupta, Jyotindra, *Language Conflict and National Development: Group Politics and National Language Policy in India* (Berkeley: University of Berkeley Press, 1970).

Datta, Satyajit, 'Look, Here's a Booming Market!', *Businessworld*, 7 February 1999, p. 21.

Delanda, Manuel, 'Markets, Antimarkets and Network Economics', Paper presented at Virtual Futures (Warwick University, UK, 1996).

Deleuze, Gilles and Félix Guattari, *A Thousand Plateaus: Capitalism and Schizophrenia* (Minneapolis: University of Minnesota Press, 1987).

Desai, Ashok, 'Why Bangalore Won', *Businessworld*, 23 September 2002, p. 10.

'Dot Hatao, Desh Bachao', *Economic Times Online*, 25 April 2000.

Dreze, Jean and Amartya Sen, *India Development and Participation* (New Delhi: Oxford University Press, 2002).

Drohan, Madelaine and Alan Freeman, 'English Rules' in *Globalization and the Challenges of a New Century: A Reader*, edited by Howard D. Mehlinger Patrick O'Meara and Mathew Krain (Bloomington: Indiana University Press, 2000).

Dubey, Rajeev, 'India as a Global R&D Hub', *Businessworld*, 17 February 2003, pp. 28–37.

Dugger, Celia, 'Connecting Rural India into the World', *New York Times*, 28 May 2000.

'E-Billionaires', *New Indian Express*, 18 September 1999.

Editorial, 'A Celebration of Freedom', *Businessworld*, 10 January 2000, p. 70.

——, 'Dr Politics and Mr Economics', *Businessworld*, 4 September 2000, p. 72.

——, 'Login to Naidunomics', *Times of India*, 18 October 1999.

——, *Financial Times*, 21 February 2003.

Einhorn, Bruce, India and IT: 'Like France and Wine', *Advantage India*, 28 January 2002; available from http://www.nasscom.org/artdisplay.asp?Art_id = 1415.

Eliade, Mircea, *Yoga: Immortality and Freedom* (Princeton: Bollingen, 1969).

'Emerging-Market Indicators', *The Economist*, 17 January 2004, p. 90.

Fishman, Joshua A., 'The New Linguistic Order' in *Globalization and the Challenges of a New Century: A Reader*, edited by Howard D. Mehlinger, Patrick O'Meara and Mathew Krain (Bloomington: Indiana University Press, 2000).

Friedman, Thomas L., *The Lexus and the Olive Tree* (New York: Anchor Books, 2000).

——, 'Tom's Journal', *The Online News Hour*, 10 March 2004; available from http://www.pbs.org/newshour/newshour_index.html.

Gardels, Nathan, *The Changing Global Order: World Leaders Reflect* (Malden, Mass.: Blackwell Publishers, 1997).

Glassman, James K., 'The Dobbs Rogue Fund', *Tech Central Station* (March 2004) Available from http://www.techcentralstation.com/021704C.html.

Gokak, V.K., *English in India: Its Present and Future* (New York: Asia Publishing House, 1964).

'Growth Scenarios of IT Industries in India', *Communications of the ACM* 44:7 (2001).

Gupta, Ashish, Paper presented at the Hyderabad IT Forum, Hyderabad (22–24 January 2003).

Gupta, R.S. and Kapil Kapoor (eds), *English in India: Issues and Problems* (New Delhi: Academic Foundation, 1991).

Gupta, R.S., 'English in Post-Colonial India' in *Who's Centric Now? The Present State of Post-Colonial Englishes*, edited by Bruce Moore (Oxford: Oxford University Press, 2001).

Gupta, Rajat, Paper presented at the TiEcon, New Delhi (8 January 2003).

Hammond, Allan, 'Digitally Empowered Development', Foreign Affairs (March–April 2001). Available from http://www.foreignaffairs.org/20010301faessay4265/allen-1-hammond/digitally-empowered-development.html.

Hammond, Allen and Elizabeth Jenkins, 'Bottom-up, Digitally Empowered Development', *Information Impacts Magazine*, February 2001.

Hari, P., 'Bangalore Technopolis', *Businessworld*, 26 February 2001.

Hegel, Georg, *Philosophy of History* (New York: Dover Publications, 1956).

Herra, Sue, 'India at Outsourcing Revolution's Core', *MSNBC*, 5 November 2003; available from http://msnbc.msn.com/id/3226069/.

Hopkins, Terence K., Immanuel Wallerstein, Robert L. Bach, Christopher Chase-Dunn and Ramkrishna Mukherjee, *World-Systems Analysis*. Vol. 1, *Explorations in the World Economy: Publications of the Fernand Braudel Center* (Beverly Hills; London; New Delhi: Sage, 1982).

'How Many Online?' *Nua: The World's leading resource for Internet trends and statistics*, Available from http://www.nua.ie/surveys/how_many_online/index.html.

Huntington, Samuel P., *The Clash of Civilizations and the Remaking of World Order*. 1st Touchstone edn (New York: Touchstone, 1997).

Ifrah, Georges, *From One to Zero: A Universal History of Numbers* (New York: Viking, 1985).

——, *The Universal History of Computing: From the Abacus to the Quantum Computer* (New York: John Wiley, 2001).

——, *The Universal History of Numbers: From Prehistory to the Invention of the Computer*. Translated by E.F. Harding David Bellos, Sophie Wood and Ian Monk (New York: John Wiley & Sons, Inc., 2000).

'India Versus China: How to Bridge the Gap', *Businessworld*, 10 June 2002, pp. 12–15.

Iyengar, Partha, Paper presented at the Hyderabad IT Forum, Hyderabad (22–24 January 2003).

Kachru, Braj, *The Alchemy of English: The Spread, Functions, and Models of Non-Native Englishes*. 1st edn, *English in the International Context* (Oxford; New York: Pergamon Institute of English, 1986).

——, *The Indianization of English: The English Language in India* (Delhi; New York: Oxford, 1983).

——, *The Other Tongue: English across Cultures* (Urbana: University of Illinois Press, 1982).

Kamdar, Mira, 'Reinventing India', *the-south-asian.com* (2001); available from www.the-south-asian.com.

Kanavi, Shivanand, 'Power Lobbying', *Business India*, 19 February to 4 March 2001, pp. 50–56.

Katakam, Anupama, 'The Warana Experiment', *Frontline*, 5–18 January 2002.

Kelly, Kevin, *Out of Control: The New Biology of Machines, Social Systems and the Economic World* (Cambridge: Perseus Books, 1994).

Khan, Nasima H., 'Diaspora Rediscovered', *India Today*, 13 May 2002, pp. 23–30.

Khanna, Rakesh, 'TARAhaat: Achieving Connectivity for the Poor Case Study', *Digital Partners*. Available from http://www.digitalpartners.org/tara.html.

Khosla, Vinod, Paper presented at the TiEcon, Silicon Valley (14–15 June 2002).

Konrad, Rachel, *Chasing the Dream*, News.Com (13 August 2001); available from http://news.com.com/2009-1017_3-270647.html.

Kotkin, Joel, *Tribes: How Race, Religion, and Identity Determine Success in the New Global Economy*, 1st edn (New York: Random House, 1993).

Kripalani, Manjeet, Pete Engardio and Steve Hamm, 'The Rise of India', *Businessweek*, 8 December 2003.

Lateef, Asma, 'Linking up with the Global Economy: A Case Study of the Bangalore Software Industry', *International Institute for Labour Studies* (1996–1997).

'Let It Shine', *The Economist*, 21 February 2004, p. 11.

Lever-Tracy, Constance, David Fu-Keung Ip and Noel Tracy, *The Chinese Diaspora and Mainland China: An Emerging Economic Synergy* (Houndmills; Basingstoke; Hampshire; New York: Macmillan Press; St. Martins Press, 1996).

Light, Ivan, 'Immigrant and Ethnic Enterprise in North America', *Ethnic and Racial Studies* 7: 2 (1984), pp. 196–216.

Light, Ivan and Steven J. Gold, *Ethnic Economies* (San Diego: Academic Press, 2000).

MacNeil, Robert, Robert McCrum and William Crun, *The Story of English* (London: Faber & Faber and BBC Books, 1992).

'Magnates in Manacles', *Businessworld*, 21 May 1999.

Mahajan, Pramod, Paper presented at the Hyderabad IT Forum, Hyderabad (22–24 January 2003).

'Man with a Mission: Interview with D. Mehta', *Financial Times*, 1999.

Manish Khanduri, Alam Srinivas, 'Software Hard Knocks', *Businessworld*, 26 March 2001, pp. 32–40.

Marx, Karl, *A Contribution to the Critique of Political Economy*. Translated by S.W. Ryazanskay (Moscow: Progress Publishers, 1970).

——, *Capital*. Translated by Eden and Cedar Paul (London: George Allen & Unwin Ltd, 1928).

——, *Grundrisse. Foundations of the Critique of Political Economy* (New York: Vintage Books, 1973).

——, 'The Future Results of British Rule in India', *NYDT*, 8 August 1853.

Marx, Karl and Frederick Engels, *The Essentials of Marx: The Communist Manifesto* (New York: Vanguard Press, 1926).

McLeish, John, *Number*, 1st American edn (New York: Fawcett Columbine, 1992).

Melwani, Lavina, *Back2Bangalore*, www.littleindia.com (9 March 2001); available from http://www.littleindia.com/India/July2001/B2B.htm.

Milton, John, *Paradise Lost and Paradise Regained, The Signet Classic Poetry Series* (New York: New American Library Inc., 1968).

'Mirage, or Reality?' *20 Years: 1982–2002. Dataquest: Collectors Issue*, 15 December 2002, pp. 100–111.

Mishra, Amit, 'Genesis of a Software Development Center', *Silicon India* (July 1999).

Murthy, Narayana, 'It Is All in the Mind, Stupid', *Business Today*, 19 January 2003, p. 160.

——, Paper presented at the Nasscom, India Leadership Forum, Mumbai (11–14 February 2003).

Naidu, B.V., 'Growth of Bangalore Software Export Industry: Role of STPI' (Bangalore: STPI, 1999).

Naroola, Gurmeet, *The Entrepreneurial Connection: East Meets West in Silicon Valley* (California USA: Special TiE Edition, 2001).

Nasscom, 'The Software Industry in India: A Strategic Review' (New Delhi: Nasscom, 1999).

Olins, Wally, *Trading Identities: Why Countries and Companies Are Taking on Each Others' Roles* (London: The Foreign Policy Centre, 1999).

'Outsourcing to India: Back Office to the World', *The Economist*, 5 May 2001, pp. 61–63.

Paul, Baran, A. and E.J. Hobsbawm, 'The Stages of Economic Growth: A Review', in *The Political Economy of Development and Underdevelopment*, edited by Charles, K. Wilber (New York: Random House, 1973).

Pecoud, Antoine, 'Thinking and Rethinking Ethnic Economies', *Diaspora* 9: 3 (2000).

Pennycook, Alastair, *The Cultural Politics of English as an International Language* (London: Longman Group Limited, 1994).

Pink, Daniel H., 'The New Face of the Silicon Age: How India Became the Capital of the Computing Revolution', *Wired* (12: 02 February 2004).

Pitroda, Sam, Paper presented at the TiEcon, New Delhi (8 January 2003).

Plato, 'Meno' in *Plato: The Collected Dialogues*, edited by Edith Hamilton and Huntington Cairns (Princeton: Princeton University Press, 1961).

'Power Lobbying' *Business India*, 19 February to 4 March 2001, pp. 52–56.

Prahalad, C.K., 'Interview: Convert India's Problems into Opportunities', *Businessworld*, 25 November 2002, pp. 38–42.

——, *India as a Source of Innovations: The First Lalbahadur Shastri National Award for Excellence in Public Administration and Management Sciences Lecture, September 30, 2000, New Delhi*, Digital Dividend (2000). Available from www.digitaldividend.org/pdf/0203ar03.pdf.

——, Paper presented at the TiEcon, New Delhi (8 January 2003).

Prahalad, C.K. and Stuart Hart, 'The Fortune at the Bottom of the Pyramid', *Digital Dividend* (2002). Available from http://www.digitaldividend.org/pubs/pubs.htm.

Proclus and Glenn R. Morrow, *A Commentary on the First Book of Euclid's Elements* (Princeton, NJ: Princeton University Press, 1970).

'Q&A with Ajit Balakrishnan, Chairman, Founder and CEO of Rediff on the Net', *Asia Source* (13 September 1999).

Rajadhyaksha, Niranjan, 'Culture Matters', *Businessworld*, 19 March 2001, p. 10.

——, 'The Chinese Opportunity', *Businessworld*, 17 February 2003, p. 8.

Rajora, Rajesh, *Bridging the Digital Divide: Gyandoot the Model for Community Networks* (New Delhi: Tata McGraw-Hill, 2002).

Raju, Ramalinga, Paper presented at the Nasscom India Leadership Forum, Mumbai (11–14 February 2003).

Ram, Mehrotra Raja, *Indian English Texts and Interpretations* (Amsterdam; Philadelphia: John Benjamins Publishing Company, 1998).

Ram, Tulsi, *Trading in Language: The Story of English in India* (New Delhi: GDK Publications, 1983).

'Ramp-up Mode', *Business Today*, 23 June 2002, pp. 59–60.

Rao, Madanmohan, *The Asia-Pacific Internet Handbook. Episode IV: Emerging Powerhouses* (New Delhi: Tata McGraw-Hill, 2002).

'Relocating the Back Office', *The Economist*, 13 December 2003, pp. 65–67.

Rekhi, Kanwal, 'President's Message', Paper presented at the TiEcon, Silicon Valley (1999).

——, 'Silicon Raj', *Upside Today*, 28 November 2000.

——, Paper presented at the TiE Speakers Meet, New Delhi (18 August 2001).

——, Paper presented at the TiEcon, Silicon Valley (14–15 June 2002).

Reynolds, Glenn, *Outsourcing Elections*, Tech Central Station (2003). Available from http://www.techcentralstation.com/062503A.html.

Richter, William, 'The Politics of Language in India', PhD Dissertation (University of Chicago, 1968).

Rostow, W.W., *The Stages of Economic Growth: A Non-Communist Manifesto* (Cambridge: Cambridge University Press, 1960).

Roy, Prasanto K., 'Foundation and Empire', *20 Years: 1982–2002. Dataquest Collectors Issue*, 15 December 2002, pp. 26–32.

Sarin, Ritu, 'Software Superpower: India Banks on the Knowledge Trade', *Asia Week*, 7 April 2000.

Satyanarayana, J., Paper presented at the Hyderabad IT Forum, Hyderabad (22–24 January 2003).

Saxenian, AnnaLee, 'Local and Global Networks of Immigrant Professionals in Silicon Valley' (Public Policy Institute of California, 2002).

——, *Regional Advantage: Culture and Competition in Silicon Valley and Route 128* (Boston: Harvard University Press, 1996).

——, *Silicon Valley's New Immigrant Experience* (California: Public Policy Institute of California, 1999).

——, 'The Bangalore Boom: From Brain Drain to Brain Circulation?' edited by Kenneth Kenniston and Deepak Kumar (Bangalore: National Institute of Advanced Study, 2000).

Schumpeter, Joseph Alois, *Capitalism, Socialism, and Democracy.* 2nd edn (New York; London: Harper & Brothers, 1947).

——, *Ten Great Economists, from Marx to Keynes* (New York: Oxford University Press, 1951).

Seife, Charles, *Zero: The Biography of a Dangerous Idea* (New York: Viking, 2000).

'Seminar on Y2K: The Road Ahead', Paper presented at the Nasscom-MAIT-CII-IT Asia (New Delhi, 1999).

Shah, R.S., Paper presented at the TiEcon, New Delhi (8 January 2003).

Shanmugaratnam, Tharman, Paper presented at the Hyderabad IT Forum, Hyderabad (22–24 January 2003).

Shekar, Meenu, 'The Changing Face of Bangalore', *Businessworld*, 22 July 1997, pp. 88–95.

Singh, Shelley, 'BPO. Cover Story', *Businessworld*, 19 August 2002, pp. 28–33.

Singh, Shelley and Mitu Jayashankar, 'The BPO Boom', *Businessworld*, 14 January 2002, pp. 28–37.

Singh, Tavleen, 'A Freedom Foiled', *India Today*, 27 May 2002, p. 24.

Singhal, Arvind and Everett M. Rogers, *India's Information Revolution* (New Delhi; Newbury Park, Calif.: Sage Publications, 1989).

——, *India's Communication Revolution: From Bullock Carts to Cyber Marts* (New Delhi: Sage, 2001).

Sircar, Arjya, 'Indianization of English Language and Literature', in *The Struggle with an Alien Tongue: The Glory and the Grief: Indianisation of English Language and Literature*, edited by R.S. Pathak (New Delhi: Babri Publications).

Slater, Joanna, 'In the Zone', *Far Eastern Economic Review*, 8 May 2003, p. 21.

Sontag, Susan, 'The World as India', *Times Literary Supplement*, 6 June 2003.

Soto, Hernando de, *The Mystery of Capital: Why Capitalism Triumphs in the West and Fails Everywhere Else* (New York: Basic Books, 2000).

Spaeth, Anthony, 'The Golden Diaspora', *Time*, 19 June 2000.

Srinivas, Alam, 'A Debugged Operating $ystem', *Outlook*, 27 January 2003.

Srivanas, Tulasi, 'A Tryst with Destiny: The Indian Case of Cultural Globalization' in *Many Globalizations: Cultural Diversity in the Contemporary World*, edited by Samuel P. Huntington and Peter L. Berger (New York: Oxford University Press, 2003), pp. 89–116.

'State of the Mart: Annual Round-up of the IT Industry', *Computers Today: Special Issue*, July 1999.

'Stolen Jobs?' *The Economist*, 13 December 2003, pp. 14–15.

Strevens, Peter, 'The Localized Forms of English', in *The Other Tongue: English across Cultures*, edited by Braj B. Kachru (Urbana: University of Illinois Press, 1982).

Sundarajan, Aruna, Paper presented at the Hyderabad IT Forum, Hyderabad (22–24 January 2003).

Tapscott, Don, Paper presented at the Nasscom India Leadership Forum, Mumbai (11–14 February 2003).

'The Great Hollowing-out Myth', *The Economist*, 21 February 2004, pp. 33–35.

'The Internet Revolution' (New Delhi: Nasscom CII-IT Asia, 1999).

'The New Jobs Migration', *The Economist*, 21 February 2004, p. 12.

'The Software Industry in India: A Strategic Review', *Nasscom* (New Delhi: Nasscom, 1999).

'The USA Gains', *Economic Times*, 2001; available from http://economictimes.com/cgi-bin/printmdd.cgi.

'The Wiring of India', *The Economist*, 25 May 2000.

'The World in 1998', *Economist Publications* (London, 1998).

Trivedi, Harish, 'Cyber-Coolies, Hindi and English; Letter', *Times Literary Supplement*, 27 June 2003.

——, 'Cyber-Coolies, Hindi and English', *Times Literary Supplement*, 22 August 2003.

Vasudev, Shefalee, 'Love, Sex and SMS', *India Today*, 14 October 2002, pp. 36–41.

Viswanathan, Vidya, 'Indian Brain Gain', *Businessworld*, 21–27 June 1999.

——, 'Indian Internet Mafia', *Businessworld*, 24 May 1999, p. 34.

——, 'The Smiles Are Back', *Businessworld*, 9 December 2002, pp. 30–32.

Wallerstein, Immanuel Maurice, *Geopolitics and Geoculture: Essays on the Changing World-System* (Cambridge [England]; New York; Paris: Cambridge University Press; Editions de la Maison des Sciences de l'Homme, 1991).

——, *The Capitalist World-Economy: Essays, Studies in Modern Capitalism* (Cambridge [England]; New York: Cambridge University Press, 1979).

——, *The Modern World-System, Studies in Social Discontinuity* (New York: Academic Press, 1974).

Wallraff, Barbara, 'What Global Language', *Atlantic Monthly*, November 2000, pp. 52–66.

Warner, Melanie, 'The Indians of Silicon Valley', *Fortune Magazine*, 15 May 2000.

Weber, Max, *From Max Weber: Essays in Sociology*. Edited by H.H. Gerth and C. Wright Mills (New York: Oxford University Press, 1946).

——, *Max Weber on Capitalism, Bureaucracy and Religion*. Edited by Stanislav Andreski (London: George Allen & Unwin, 1983).

——, *The Protestant Ethic and the Spirit of Capitalism*. Translated by Talcott Parsons (New York: Charles Scribner's Sons, 1958).

'What's Stopping Us? Businessworld Round Table', *Businessworld*, 17 February 2003, pp. 20–26.

Whitmore, Stuart, 'Driving Ambition', *Asiaweek*; available at http://www.asiaweek.com/asiaweek/technology/990625/bhatia_2.html.

Yergin, Daniel and Joseph Stanislaw, *The Commanding Heights: The Battle for the World Economy* (New York: Simon & Schuster, 2002).

Yergin, Daniel and William Cran, 'The Agony of Reform' in *Commanding Heights: The Battle for the World Economy* (WGBH, 2003).

Yergin, Daniel and William Cran, 'The New Rules of the Game' in *Commanding Heights: The Battle for the World Economy* (WGBH, 2003).

'Yes, Yes – 2000 Times Yes!', *The Economist*, 4 October 1997.

Yourdon, Edward, *Decline & Fall of the American Programmer*, Yourdon Press Computing Series (Englewood Cliffs, NJ: Yourdon Press, 1992).

Zwingle, Erla, 'World Cities', *National Geographic*, November 2002, pp. 75–99.

Cited interviews

Acharya, Ranjan and Anurag Behar, *Wipro* (Bangalore, 20 January 2003).

Cravalaho, Brian, *ICICI OneSource* (Bangalore, 21 January 2003).

Joshi, Kailash, *TiE* (Silicon Valley, 17 June 2002).
Manzar, Osama, *Inomy* (New Delhi, 10 January 2003).
Mehta, Sunil, *Nasscom* (New Delhi, 13 January 2003).
Mishra, Vish, *TiE* (Silicon Valley, 17 June 2002).
Patnaik, Manas, *STPI* (Bhubaneshwar, June 1999).
Patwardhan, Anand, *IIT* (Mumbai, 6 February 2003).
Rao, Madanmohan, *Inomy* (Bangalore, 18 January 2003).
Rekhi, Kanwal, *TiE* (Silicon Valley, 17 June 2002).
Sen, Ashish, *Voices* (Bangalore, 20 January 2003).
Thakur, Animesh, *Hero Mindmine* (Mumbai, 4 February 2003).
Walters, Ravindra, *NeoIT* (Bangalore, 17 January 2003).

Index

Printed in the United States
50919LVS00001B/307-309

9 781403 939432